PRAISE FOR *GUSHER OF LIES*

"A must-read for anyone interested in energy issues. Concerning the topic of energy and the many myths associated with energy issues, this is a debunker's bible." —R-Squared Energy Blog

"Bryce does a fantastic job of helping people understand the sheer magnitude of energy flows that would have to be replaced to attain energy independence, and conclusively makes his case that pursuing energy interdependence is a superior objective."

—*New York Post*

"The higher oil prices go—and oil company profits with them—the greater the temptation to seek miracle cures for our energy problems. Mr. Bryce reminds us that as important as energy is, it does not stand apart from a national economy that is deeply connected to the rest of the world, any more than it can be divorced from the laws of thermo-dynamics. Nor should his informed skepticism be mistaken for cynicism or a sense of futility. His realistic portrayal of our energy situation is timely and important, dismissing widespread notions of quick-and-easy solutions and making a strong case that the current yearning for energy self-sufficiency, while understandable, is both unattainable and inconsistent with the basis of much of our post-World War II success."

—*Energy Outlook*

"[C]arefully, gleefully throttles the meaningless rhetoric driving the cry for energy independence. . . . High-order muckraking and an excellent primer for addressing the real question: How are we going to handle energy interdependence?" —*Kirkus Reviews*

"Energy is a very strange business. It takes a brave soul to step into the political fray and make straight calls on facts, merits and the hidden and not-so-hidden agendas of many of the players. Robert Bryce's lucid take in *Gusher of Lies* on global energy realities is both engaging and compelling."

—MARK P. MILLS, writer of the *Forbes* column "Energy Intelligence," and co-author of *The Bottomless Well: The Twilight of Fuel, the Virtue of Waste, and Why We Will Never Run Out of Energy*

"In *Gusher of Lies* Robert Bryce does political leaders around the world an enormous favor by debunking in its entirety the myth that anymajor energy-consuming countrywith flat or declining energy-supplies can ever achieve the utopia called 'energy independence.' He also lucidly spells outexactly whyAmerica is the least likely country even to come close."

—MATTHEW SIMMONS, chairman of the Houston-based investment banking firm Simmons & Company International, and author of *Twilight in the Desert: The Coming Saudi Oil Shock and the World Economy*

"He blasts Republicans, Democrats, the presidential candidates, Al Gore, Robert Redford, environmentalists, and energy analysts for misleading the public about our energy needs. . . . Meticulously re-searched with copious facts—nearly all footnoted—this illuminating and sometimes witty work offers another view of the current state of energy."   —*Library Journal*

"Veteran energy analyst Robert Bryce challenges what has become a policy axiom of the American political establishment . . . [and] demol-ishes the many 'false promises' that are promoted by those calling for energy independence."   —*Business Times Singapore*

"Ever since Richard Nixon's dreams of the early 1970s, energy inde-pendence has been promoted as both desirable and possible, and for different reasons, it has found advocates right across the country's political spectrum. Over the decades many sensible observers have remarked on the delusionary nature of this goal. But in this book Robert Bryce goes a crucial step further as he systematically demolishes this persistent myth by using a wide range of readily available statistics and consistent arguments to deconstruct misinformation about the nature of America' energy systems and to straighten the misunderstandings about the workings of global energy markets."

—VACLAV SMIL, distinguished professor, University of Manitoba, and author of numerous books including *Energy: A Beginner's Guide* and *Energy at the Crossroads: Global Perspectives and Uncertainties*

"Bryce handily dispels the public's misconceptions about energy. . . . He's spot-on."   —*San Antonio Express-News*

"Bryce methodically exposes . . . mistaken assumptions and bad logic. . . . He presents to the reader a strong set of arguments to prove that energy independence is impossible (unless we want to turn the clock way, way back) and that all of the proposed means for achieving it are certain to fall short of their over-hyped expectations."

—*Regulation* Magazine

"Bryce's new book couldn't be more timely. . . . Fascinating."

—*Tucson Citizen*

GUSHER OF LIES

GUSHER OF LIES

The Dangerous Delusions of "Energy Independence"

ROBERT BRYCE

PublicAffairs
New York

Published in the United States by PublicAffairs™,
a member of the Perseus Books Group.

PublicAffairs books are available at special discounts for bulk
purchases in the U.S. by corporations, institutions, and other organizations.
For more information, please contact the Special Markets Department at
the Perseus Books Group, 2300 Chestnut Street, Suite 200, Philadelphia, PA 19103,
call (800) 810-4145, ext. 5000, or e-mail special.markets@perseusbooks.com.

Text design by Trish Wilkinson

Library of Congress Cataloging-in-Publication Data

Bryce, Robert.
 Gusher of lies : the dangerous delusions of energy independence / Robert
Bryce.
     p.   cm.
 Includes bibliographical references and index.
 HC ISBN 978-1-58648-321-0 (hard cover : alk. paper)
 PB ISBN 978-1-58648-690-7
 1. Power resources—United States. 2. Energy policy—United States. I. Title.
TJ163.25.U6B79 2008
333.790973—dc22   2007041459

10 9 8 7 6 5 4 3 2 1

What are the facts? Again and again and again—what are the facts? Shun wishful thinking, ignore divine revelation, forget what "the stars foretell," avoid opinion, care not what the neighbors think, never mind the unguessable "verdict of history"—what are the facts, and to how many decimal places? You pilot always into an unknown future; facts are your single clue. Get the facts!

Robert Heinlein, 1907–1988

CONTENTS

PART ONE

WHY WE THINK WE
WANT ENERGY INDEPENDENCE

PART TWO

FROM DOMINANCE TO DEPENDENCE:
AMERICAN ENERGY HISTORY,
RHETORIC, AND THE NEW REALITIES

LIST OF FIGURES
AND TABLES

FIGURES

TABLES

AUTHOR'S NOTE

The concept for this book grew out of my first two books. In the course of writing *Pipe Dreams* and *Cronies,* I began to understand the multitude of ways that the energy business influences local, national, and global politics. Those books, and this one, are part of my ongoing fascination with the energy business. From the financial to the technological, from the drill bit to the spark plug, the energy sector is the biggest, most interesting, most dynamic, most important business on the planet. It's also the most misunderstood.

I did not write this book with a political agenda—at least not one that comes from any partisan convictions. I am neither Democrat nor Republican. I am a charter member of the Disgusted Party. I'm a radical centrist, a raging moderate, who leans toward the libertarian and believes wholeheartedly in the U.S. Constitution and the Bill of Rights.

Although I wrote this book with the hope that it will bring a smidgen of sensibility to the debate over energy, I am not overly sanguine about the prospects. Neither the Republicans nor the Democrats appear serious about addressing America's energy needs. And therein lies the key problem in America's energy discussions: There's far too much religion and far too little science. Politicians and pundits prefer demagoguery to the reality of the global energy marketplace. This book was written as a rebuke to that partisanship and as a reminder to Americans and energy consumers everywhere that they are, like it or not, participants in a global economy—one that trades in everything from gasoline to fresh flowers—and they must accept that fact.

This book reflects an ongoing shift in my views on energy policy. Three or four years ago, I bought into some of the concepts that I debunk in this book. I used to think that a government-sponsored "Manhattan Project" for energy development was the key to resolving America's future energy needs. I have advocated a carbon tax. I have also written essays that accepted the neoconservatives' claims that America's oil needs are directly related to terrorism.

I no longer subscribe to those notions. Instead, I am increasingly of the view that government needs to quit meddling in the energy market for a simple reason: Each time Congress or the White House gets too involved in the energy business, supplies get tighter or prices increase, or both. Of course, politicians always want to "help." And the energy sector provides a perfect venue for demagogues to bash the evils of Big Oil, or Big Coal, or Bad Arabs, or Evil OPEC. Unfortunately for consumers, history has repeatedly shown that congressional meddling usually ends in a muddle. It also shows that the less regulation imposed on the energy sector, the better, usually, for consumers.

On a more personal note, this book had several fits and starts and took far longer to write than it probably should have.

Many thanks to A. F. Alhajji at Ohio Northern University. He facilitated my 2006 visit to Saudi Arabia and the United Arab Emirates—a trip that helped me begin to understand the Arab world and the ongoing trend of globalization within the energy sector. He has also been a wonderful sounding board and friend. Thanks to Abdulwahab al-Faiz, the editor of *Al Eqtisadiah,* the Saudi business newspaper, for his encouragement and his desire to have my opinions published in Arabic.

Mike Ameen offered a wealth of historical insights, encouragement, and a Rolodex that includes contacts around the world. (Better still, he never lets me pick up the restaurant bill.) Thanks also to my pal Robert Elder, Jr., a terrific reporter who offered numerous insights and many helpful suggestions on the various drafts of this manuscript. Special thanks to my favorite father-in-law and favorite chemist, Paul G. Rasmussen, for his patience and guidance on matters of thermodynamics and ethanol. He patiently read many drafts and offered helpful tweaks and ideas about style. My understanding of ethanol was greatly enhanced by my many conversations with my friend Tad Patzek of the University of California at Berkeley, who offered technical help and encouragement. Critical help on the ethanol front came from Jan Kreider and Bill Reinert, who alerted me to the huge water costs associated with large-scale ethanol production.

Thanks, too, to Mark Mills, Donald Stedman, Vaclav Smil, Guy Caruso, Arthur L. Smith, John S. Herold Inc., Tom Morehouse, Abdulaziz Sager, the Gulf Research Center, Mauro Renteria, Ron and Violet Cauthon, Bryan Shahan, Yazan Rahman, Steve Miller, Adnan Suboh, Matthew Simmons, Will van Overbeek, Charley Maxwell, Jim Moore, H. J. Gruy, Jay Tapp, Thomas Palaima, Ross Milloy, Farris Rookstool III, Khaled bu-Ali, Maan al-Saena, and Salah Tamari. In addition, I want to thank Chuck Spinney, Greg Wilcox, G. I. Wilson, Allen Gill, Rex Rivolo, and Donald Vandergriff, all of whom offered encouragement and advice throughout my research and writing. Key help came from Robert L. Bradley, Jr., the president of the Institute for Energy Research, who offered many insightful suggestions and politely made it clear where I was being careless in my thinking and writing.

Omar Kader was a great help, offering patient advice and a substantial amount of time. He reminded me to stay focused—and he did so at a critical time in the writing of this book. Peter Wells and Sarah Lloyd were a tremendous help in providing contacts in Dubai and Kuwait. So, too, was Khaled al-Shaya, who introduced me to a number of helpful people in the Kuwaiti oil sector. I must also tip my hat to Alex Economides and Jay Clark of the *Energy Tribune,* who helped with various charts and data. In particular, I must thank the magazine's editor, Michael J. Economides, who has been supportive in a number of key ways.

I got outstanding research help from Les McLain, an Excel whiz who helped manage a myriad of spreadsheets, charts, and minutiae.

She made this book better. My friend and fact checker, Mimi Bardagjy, not only caught errors, but she also offered savvy advice on editing and organization. While all of these people provided assistance, any errors in these pages are my own.

I must acknowledge one bit of software that made this book possible: NoteTaker. I have used dozens of software programs, but Note-Taker, made by AquaMinds, is one of the most useful pieces of software I've come across.

I need to thank my agent, Dan Green, for his patience and gentle guidance. Thanks to my favorite editor on the planet, Lisa Kaufman at PublicAffairs, who continues to show her faith in me. She, Susan Weinberg, and Peter Osnos remained patient and encouraging as I struggled to find the book within the material I was collecting.

A thousand thanks to my children, Mary, Michael, and Jacob, for their patience and love. All of them are insanely great. And finally, there are no words capable of expressing my affection and appreciation to my wife, Lorin—my first reader, my first editor, and my one, true love.

INTRODUCTION

The Persistent Delusion

Americans love independence.

Whether it's financial independence, political independence, the Declaration of Independence, or grilling hotdogs on Independence Day, America's self-image is inextricably bound to the concepts of freedom and autonomy. The promises laid out by the Declaration—life, liberty, and the pursuit of happiness—are the shared faith and birthright of all Americans.

Alas, the Founding Fathers didn't write much about gasoline.

Nevertheless, over the past 30 years or so—and particularly over the past 3 or 4 years—American politicians have been talking as though Thomas Jefferson himself warned about the dangers of imported crude oil. Every U.S. president since Richard Nixon has extolled the need for energy independence. In 1974, Nixon promised it could be achieved within 6 years.[1] In 1975, Gerald Ford promised it in 10.[2] In 1977, Jimmy Carter warned Americans that the world's supply of oil would begin running out within a decade or so and that the energy crisis that was then facing America was "the moral equivalent of war."[3]

The phrase "energy independence" has become a prized bit of meaningful-sounding rhetoric that can be tossed out by candidates and political operatives eager to appeal to the broadest cross section of

1

voters. When the U.S. achieves energy independence, goes the reasoning, America will be a self-sufficient Valhalla, with lots of good-paying manufacturing jobs that will come from producing new energy technologies. Farmers will grow fat, rich, and happy by growing acre upon acre of corn and other plants that can be turned into billions of gallons of oil-replacing ethanol. When America arrives at the promised land of milk, honey, and supercheap motor fuel, then U.S. soldiers will never again need visit the Persian Gulf, except, perhaps, on vacation. With energy independence, America can finally dictate terms to those rascally Arab sheikhs from troublesome countries. Energy independence will mean a thriving economy, a positive balance of trade, and a stronger, better America.

The appeal of this vision of energy autarky has grown dramatically since the terrorist attacks of September 11. That can be seen through an analysis of news stories that contain the phrase "energy independence." In 2000, the Factiva news database had just 449 stories containing that phrase. In 2001, there were 1,118 stories. By 2006, that number had soared to 8,069.

The surging interest in energy independence can be explained, at least in part, by the fact that in the post–September 11 world, many Americans have been hypnotized by the conflation of two issues: oil and terrorism. America was attacked, goes this line of reasoning, because it has too high a profile in the parts of the world where oil and Islamic extremism are abundant. And buying oil from the countries of the Persian Gulf stuffs petrodollars straight into the pockets of terrorists like Mohammad Atta and the 18 other hijackers who committed mass murder on September 11.

Americans have, it appears, swallowed the notion that all foreign oil—and thus, presumably, all foreign energy—is bad. Foreign energy is a danger to the economy, a danger to America's national security, a major source of funding for terrorism, and, well, just not very patriotic. Given these many assumptions, the common wisdom is to seek the balm of energy independence. And that balm is being peddled by the Right, the Left, the Greens, Big Agriculture, Big Labor, Republicans,

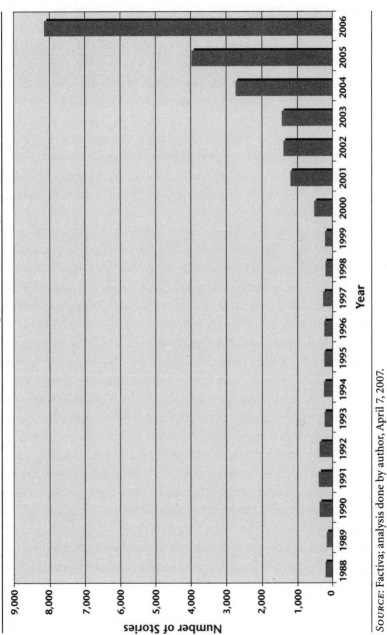

FIGURE 1 Stories in the Factiva News Database Containing the Phrase "Energy Independence," 1988–2006

SOURCE: Factiva; analysis done by author, April 7, 2007.

Democrats, senators, members of the House, George W. Bush, the opinion page of the *New York Times,* and the neoconservatives. About the only faction that dismisses the concept is Big Oil. But then few people are listening to Big Oil these days.

Environmental groups like Greenpeace and Worldwatch Institute continually tout energy independence.[4] The idea has long been a main talking point of Amory Lovins, the high priest of the energy-efficiency movement and the CEO of the Rocky Mountain Institute.[5] One group, the Apollo Alliance, which represents labor unions, environmentalists, and other left-leaning groups, says that one of its primary goals is "to achieve sustainable American energy independence within a decade."[6]

Al Gore's 2006 documentary about global warming, *An Inconvenient Truth,* implies that America's dependence on foreign oil is a factor in global warming.[7] The film, which won two Academy Awards (for best documentary feature and best original song), contends that foreign oil should be replaced with domestically produced ethanol and that this replacement will reduce greenhouse gases.[8] (In October 2007, Gore was awarded the Nobel Peace Prize.)

The leading Democratic candidates for the White House in 2008 have made energy independence a prominent element of their stump speeches. Illinois senator Barack Obama has declared that "now is the time for serious leadership to get us started down the path of energy independence."[9] In January 2007, in the video that she posted on her Web site that kicked off her presidential campaign, New York senator Hillary Clinton said she wants to make America "energy independent and free of foreign oil."[10] Former North Carolina senator John Edwards believes the U.S. needs "energy independence from unstable and hostile areas of the world."[11]

The Republicans are on board, too. In January 2007, shortly before Bush's State of the Union speech, one White House adviser declared that the president would soon deliver "headlines above the fold that will knock your socks off in terms of our commitment to energy independence."[12] In February 2007, Arizona senator and presidential candidate John McCain told voters in Iowa, "We need energy independence. We need it for a whole variety of reasons."[13] In March 2007, former New York mayor Rudolph Giuliani insisted that the federal government "must treat energy independence as a matter of national security." He went on, saying that "we've been talking about energy independence for over 30 years and it's been, well, really, too much talk and virtually no action. . . . I'm impatient and I'm single-minded about my goals, and we will achieve energy independence."[14]

On April 26, 2007, another Republican candidate, Mitt Romney, the former governor of Massachusetts, used the *Jerusalem Post*'s e-mail list to conflate the issues of oil, terrorism, Israel, and energy independence in a fund-raising appeal for his presidential campaign. The e-mail message, which showed a large picture of Iranian president Mahmoud Ahmadinejad, asked several questions, including "Do you believe that those who support terrorism against America and against the state of Israel should be held accountable?" The next question: "Do you agree that we must become energy independent and stop sending $1 billion a day to nations like Iran and Syria who use that money against us?"[15] (Syria exports modest amounts of crude oil.[16])

The Democratic Party, which won control of the House and Senate in the November 2006 elections, has made energy independence a key talking point. About the time of the elections, Nancy Pelosi, the congresswoman from San Francisco who became Speaker of the House, issued the Democrats' "New Direction" agenda. The third point on that list—right after raising the minimum wage and repealing certain tax incentives—is "invest in research and development to promote energy independence." It says the Democrats will achieve energy independence "within ten years. We should be sending our energy dollars to the Midwest, not the Middle East. America's farmers will fuel America's energy independence."[17]

A Democratic think tank, the Center for American Progress, which was created by a group of politicos from the Clinton administration, has launched a campaign called "Kick the Oil Habit," an effort that seems to imply America can quit using oil with the same ease that a smoker might give up cigarettes.[18] In May 2006, the group's lead spokesman, actor Robert Redford, appeared on TV talk shows and wrote opinion pieces in which he said the U.S. should quit using oil altogether so that it can get away from "dictators and despots." The solutions proposed by Redford and the Democrats: more ethanol, biofuels, and hybrid vehicles.[19] During an appearance on CNN's *Larry King Live,* Redford said that he supported corn ethanol production because "it's cheaper. It's cleaner. It's renewable. And you know what? It's American because we grow it."[20]

In January 2007, Andy Grove, the former chairman of giant computerchip maker Intel Corp., penned an opinion piece for the *Wall Street Journal* in which he decried the lack of progress toward energy independence: "Even though the importance of the energy independence issue has been recognized and emphasized by every president since 1974, our vital national objective is vanishing like a mirage in the distance." Grove went on to claim that our use of foreign energy "gives great power to other nations over our destiny."[21]

In September 2007, S. David Freeman, a longtime advocate of renewable energy who once chaired the Tennessee Valley Authority and has headed several other electric utilities, released a book called *Winning Our Energy Independence: An Energy Insider Shows How.* Freeman's book calls for a multidecade effort to close America's older coal and nuclear power plants while focusing on more efficient plug-in hybrid cars. A press release publicizing the book says that "Freeman charges that the reason we aren't already using more renewable energy is that the oil companies and electrical utilities have waged a slick campaign to deceive Americans."[22]

In October 2007, a book with a similar theme—*Freedom from Oil: How the Next President Can End the United States' Oil Addiction*—rose to number 8 on the *Washington Post's* bestseller list. The book, by David Sandalow, a senior fellow at the Brookings Institution and a former official in the Clinton administration, touts the potential of plug-in hybrid cars, biofuels, and fuel efficiency to cut America's oil consumption. The front cover of the book has a blurb from Al Gore which says that when Sandalow "writes about energy and the environment, we should all pay close attention."

Polls show that an overwhelming majority of Americans are worried about foreign oil. A March 2007 survey by Yale University's Center for Environmental Law and Policy found that 93 percent of respondents said imported oil is a serious problem and 70 percent said it was "very" serious.[23] That finding was confirmed by an April 2007 poll by Zogby International, which found that 74 percent of Americans believe that cutting oil imports should be a high priority for the federal government. And a majority of those surveyed said that they support expanding the domestic production of alternative fuels.[24]

The energy independence rhetoric has become so extreme that some politicians are even claiming that lightbulbs will help achieve the goal. In early 2007, U.S. Representative Jane Harman, a California Democrat, introduced a bill that would essentially outlaw incandescent bulbs by requiring all bulbs in the U.S. to be as efficient as compact fluorescent bulbs. Writing about her proposal in the *Huffington Post,* Harman declared that such bulbs could "help transform America into an energy efficient and energy independent nation."[25]

While Harman may not be the brightest bulb in the chandelier, there's no question that the concept of energy independence resonates with American voters and explains why a large percentage of the American populace believes that energy independence is not only doable but desirable.

But here's the problem, and the reason for this book: It's not and it isn't.

Energy independence is hogwash. From nearly any standpoint—economic, military, political, or environmental—energy independence makes no sense. Worse yet, the inane obsession with the idea of energy independence is preventing the U.S. from having an honest and effective discussion about the energy challenges it now faces.

This book focuses on the need to acknowledge, and deal with, the difference between rhetoric and reality. The reality is that the world—and the energy business in particular—is becoming ever more interdependent. And this interdependence will likely only accelerate in the years to come as new supplies of fossil fuel become more difficult to find and more expensive to produce. While alternative and renewable forms of energy will make minor contributions to America's overall energy mix, they cannot provide enough new supplies to supplant the new global energy paradigm, one in which every type of fossil fuel—crude oil, natural gas, diesel fuel, gasoline, coal, and uranium—gets traded and shipped in an ever more sophisticated global market.

Regardless of the ongoing fears about oil shortages, global warming, conflict in the Persian Gulf, and terrorism, the plain, unavoidable truth is that the U.S., along with nearly every other country on the planet, is married to fossil fuels. And that fact will not change in the foreseeable future, meaning the next 30 to 50 years. That means that the U.S. and the other countries of the world will continue to need oil and gas from the Persian Gulf and other regions. Given those facts, the U.S. needs to accept the reality of *energy interdependence*.

The integration and interdependence of the $5-trillion-per-year global energy business can be seen by looking at Saudi Arabia, the biggest oil producer on the planet.[26] In 2005, the Saudis *imported* 83,000 barrels of gasoline and other refined oil products per day.[27] It can also be seen by looking at Iran, which imports 40 percent of its gasoline needs. Iran also imports large quantities of natural gas from Turkmenistan.[28] If the Saudis, with their 260 billion barrels of oil reserves, and the Iranians, with their 132 billion barrels of oil and 970 trillion cubic feet of natural gas reserves, can't be energy independent, why should the U.S. even try?[29]

An October 2006 report by the Council on Foreign Relations put it succinctly: "The voices that espouse 'energy independence' are doing the nation a disservice by focusing on a goal that is unachievable over the foreseeable future and that encourages the adoption of inefficient and counterproductive policies."[30]

America's future when it comes to energy—as well as its future in politics, trade, and the environment—lies in accepting the reality of an increasingly interdependent world. Obtaining the energy that the U.S. will need in future decades requires American politicians, diplomats, and businesspeople to be actively engaged with the energy-producing countries of the world, particularly the Arab and Islamic producers.

Obtaining the country's future energy supplies means that the U.S. must embrace the global market while acknowledging the practical limits on the ability of wind power and solar power to displace large amounts of the electricity that's now generated by fossil fuels and nuclear reactors.

The rhetoric about the need for energy independence continues largely because the American public is woefully ignorant about the fundamentals of energy and the energy business.[31] It appears that voters respond to the phrase, in part, because it has become a type of code that stands for foreign policy isolationism—the idea being that if only the U.S. didn't buy oil from the Arab and Islamic countries, then all would be better. The rhetoric of energy independence provides political cover for protectionist trade policies, which have inevitably led to ever larger subsidies for politically connected domestic energy producers, the corn ethanol industry being the most obvious example.

But going it alone with regard to energy will not provide energy security or any other type of security. Energy independence, at its root, means protectionism and isolationism, both of which are in direct opposition to America's long-term interests in the Persian Gulf and globally.

Once you move past the hype and the overblown rhetoric, there's little or no justification for the push to make America energy independent. And that's the purpose of this book: to debunk the concept of energy independence and show that none of the alternative or renewable energy sources now being hyped—corn ethanol, cellulosic ethanol, wind power, solar power, coal-to-liquids, and so on—will free America from imported fuels. America's appetite is simply too large and the global market is too sophisticated and too integrated for the U.S. to secede.

Indeed, America is getting much of the energy it needs because it can rely on the strength of an ever-more-resilient global energy market. In 2005, the U.S. bought crude oil from 41 different countries, jet fuel from 26 countries, and gasoline from 46.[32] In 2006, it imported coal from 11 different countries and natural gas from 6 others.[33] American consumers in some border states rely on electricity imported from Mexico and Canada.[34] Tens of millions of Americans get electricity from nuclear power reactors that are fueled by foreign uranium. In 2006, the U.S. imported the radioactive element from 8 different countries.[35]

Yes, America does import a lot of energy. But here's an undeniable truth: It's going to continue doing so for decades to come. Iowa farmers can turn all of their corn into ethanol, Texas and the Dakotas can cover themselves in windmills, and Montana can try to convert all of its coal into motor fuel, but none of those efforts will be enough. America needs energy, and lots of it. And the only way to get that energy is by relying on the vibrant global trade in energy commodities so that each player in that market can provide the goods and services that it is best capable of producing.

This book is designed to provide a sober look at America's energy situation. To that end, it is divided into several sections.

Part 1 examines the appeal of energy independence, a concept that has gained traction in the minds of many Americans because it appears to offer a solution to a number of thorny problems now faced by the U.S. This section details the many false promises that lie behind the rhetoric of energy independence. It also looks at America's energy imports and compares them to the imports of other key mineral commodities.

Part 2 discusses America's history in the global energy market and how it went from being a dominant producer that dictated the global price of oil to an oil importer that has the price of oil dictated to it by OPEC. This section explains how nearly every presidential administration since that of Richard Nixon has responded to this shift in power by making strategic alliances with certain Persian Gulf countries, by militarizing the region, and by promising that energy independence lay just around the nearest corner service station. It will also show how the latest push for energy independence is being led by the same group of warmongering neoconservatives who led the cheerleading for the Second Iraq War.

Part 3 provides a discussion of why the U.S. cannot wean itself off foreign energy. There are three main reasons: Energy use keeps growing; energy efficiency won't necessarily mean a reduction in consumption; and most important, renewable energy and alternative fuels simply cannot provide the volume of energy needed to replace traditional fossil fuels at any time in the foreseeable future.

This section provides a comprehensive dissection of the ethanol scam. The promise of energy independence has given powerful members of Congress the excuse they need to provide ever greater subsidies to special interests, Big Corn and Big Ethanol being the primary beneficiaries. Whether the issue is subsidies, food supplies, land use, air pollution, energy balance, Brazilian ethanol, or the way ethanol affects the selection of America's presidential candidates, ethanol is one of the biggest frauds ever perpetrated on U.S. taxpayers. In addition to providing a critical look at ethanol, this section examines the challenges facing other energy sources, including natural gas, wind, solar, coal, and nuclear power. And it shows why none of those sources will be able to provide enough energy to obviate the need for imports.

Part 4 discusses the rising power and influence of the Arab and Islamic states in the Persian Gulf and tells why the U.S. cannot ignore this trend. Saudi Arabia, Dubai, and Iran are all gaining influence, much of which is due to their energy resources. For the Saudis and the Iranians, that influence comes directly from oil and gas. Dubai's influence is coming from the emirates' skill in trading and its embrace of open markets. This section gives examples of the world's growing energy interdependence and offers ideas about how the U.S. should move forward with regard to energy over the coming years and decades.

The goal throughout this book is to use common sense and easily verifiable facts—nearly all of them footnoted—not hyperbole and emotion.

There is no partisan agenda at work in these pages. There is no such thing as Democratic kilowatt-hours or Republican gasoline. Consumers don't purchase liberal electricity or conservative motor fuel. Their interest is in obtaining the energy they need at affordable prices.

This book is designed to provide facts, not propaganda. Understanding the facts behind America's energy situation requires perspective. And that requires a deeper understanding of how America's energy imports compare to imports of other essential commodities. While American politicians are obsessed with imported oil, little attention is given to the potential dangers of America's need for platinum, even though the U.S. imports 91 percent of the platinum that it consumes. The U.S. relies on foreign suppliers for dozens of other critical commodities, ranging from semiconductors to steel.[36] All of which raises an obvious question: Why should America stop at energy independence? Why not demand fresh-flower independence? Or perhaps iPod independence?

That's the focus of Chapter 1.

PART ONE

WHY WE THINK WE WANT ENERGY INDEPENDENCE

1

IMPORTS ARE US

Oil is a strategic commodity. And it always will be. But America's oil imports must be seen in context alongside its reliance on other critical commodities like semiconductors and palladium.

A vocal group of anti-foreign-oil pundits claim that America imperils its security because of its reliance on Persian Gulf oil producers. That claim persists even though the countries of the Persian Gulf supplied just 11 percent of all the oil consumed in the U.S. in 2005.[1] Sure, that 11 percent is essential. And yes, it would be difficult for the U.S. to replace those imports. But those fuel imports are just one facet of America's overall import picture.

Although the U.S. imports about 60 percent of its total oil needs, it imports about 80 percent of its semiconductors.[2] Like oil, semiconductors are an essential commodity. And yet, the U.S. has not deployed the 82nd Airborne Division to Taiwan as an insurance policy against the possibility that a foreign power might halt the flow of flash memory, processors, and other computer hardware. Nor are any politicians declaring the need for America to be "semiconductor independent." As Ivan Eland, the savvy analyst and columnist at the Independent Institute, points out, "Oil accounts for between 65 and 95 percent of the exports of Persian Gulf nations. In contrast, oil makes up only about 7 percent of U.S. imports. Thus, most states, whether their governments are friendly to the United States or not, have a huge incentive to export oil into the world market."[3]

The oil-rich nations in the Persian Gulf and elsewhere have to sell their petroleum. They can't drink it, nor use it to water their palm trees. They must sell it. And America and other oil-importing countries are eager buyers. As with every other product or service, the buyer and the seller reach a price and the deal is done. Both buyer and seller benefit from the transaction. And the more business they do, the more interdependent they become.

That interdependence extends far beyond oil and semiconductors. America is heavily dependent on imported minerals. The U.S. imports 100 percent of its bauxite, alumina, manganese, strontium, yttrium, and 13 other strategic mineral commodities. (For the full list of the mineral commodities that the U.S. imports and their individual uses, see Appendix A.)

According to the U.S. Geological Survey, the U.S. imports 99 percent of its gallium, 91 percent of its platinum, 88 percent of its tin, 81 percent of its palladium, 76 percent of its cobalt, and 72 percent of its chromium.[4] All of those items are essential commodities in the American economy. So why hasn't Barack Obama called for palladium independence? Why doesn't George W. Bush insist on cobalt independence?

The delusion of energy independence is attracting adherents at the same time that Americans are fixated on energy costs. There is a widespread perception that oil and other forms of energy are near record highs in terms of cost. But that perception is flat wrong. Here's a startling fact: In early 2007, Americans were paying less for gasoline than they did back in 1919. The seldom-mentioned and seemingly counterintuitive truth about America's energy business is that it keeps getting better and better at delivering more and more energy and doing so at lower and lower prices. And part of the reason why American energy prices have fallen is that U.S. consumers are able to rely on the global market for oil, gas, coal, electricity, and uranium.

In 1919, gasoline (on an inflation-adjusted basis) cost about $2.97 per gallon.[5] That year, the U.S. was using about 2.7 billion gallons of motor fuel per year, or about 176,000 barrels of oil per day.[6] Today, the U.S. is using about 21 million barrels of oil per day, or about 119 times as much as it was back in 1919, and yet real motor fuel prices have *fallen*.[7] In fact, oil—along with most other energy forms that Americans buy—remains remarkably cheap by almost any measure. In early August 2007, regular gasoline in the U.S. cost, on average, $2.82 per gallon—that's $0.15 cheaper than it was back in 1919.[8] The same trend can be seen in crude oil prices. In mid-2007, crude oil prices were about the same, again, in inflation-adjusted dollars, as they were in 1983.[9]

So why have prices for gasoline and other oil products fallen (or stayed flat) over the last eight or nine decades? Well, one key reason is obvious: The U.S. can import those commodities from lots of different countries. *And the U.S. has been doing just that for nearly a century.*

The U.S. was a net crude oil importer way back in 1913. In fact, between 1913 and 2007, the U.S. was a net crude *exporter* in just nine of those years.[10] In 1913—just five years after Henry Ford began selling his Model T—America was importing 36,000 barrels of crude oil per day.[11] Nine decades later, in 2005, with George W. Bush in the White House, the U.S. was importing almost 300 times as much oil as it did when Woodrow Wilson was living at 1600 Pennsylvania Avenue.[12]

But once again, those numbers must be put in perspective. Over the past century or so, America's energy consumption has grown in direct relation to its economic growth: In 1913, America's gross domestic product was about $39 billion. By 2005, U.S. GDP was more than $12.4 trillion, or about 300 times as much as the 1913 figure.[13] Thus, in a remarkable parallel, that 300-fold increase in oil imports has been accompanied by a 300-fold increase in America's economic output.

Further, the imports from 1913 must be considered in terms of the number of motor vehicles in use. Back in 1913, there were only about 1.25 million motor vehicles in America.[14] Nine decades later, that fleet has grown nearly 200-fold to some 243 million motor vehicles.[15]

The trends in oil prices are similar to those in electric power: Electricity isn't getting more expensive, it's getting cheaper. In 2005, electricity in the U.S. (again on an inflation-adjusted basis) was about 16 percent cheaper than it was in 1960, partly because coal is getting cheaper.[16] In 2005, coal was about 10 percent cheaper than it was back in 1960.[17] Why is it cheaper? Well, in part because of technology. American miners are getting more efficient in their production techniques. Another factor: American coal producers must compete with foreign coal producers.

Many Americans don't believe energy is getting cheaper for a simple reason: They are using more of it. Consumption of both oil and electricity in the U.S. has been increasing for decades, and those increases tend to cancel out any noticeable reductions in price. Nevertheless, the clear and unavoidable truth about energy imports—as well as the myriad other items that the U.S. imports from foreign countries—is that imports are critically important to the overall health of the American economy. The U.S. cannot survive without foreign crude, nor can it manage without imports of rare minerals like gallium, which is needed to make semiconductors, photovoltaic cells, and lasers.

Over the past few years, the essentiality of imports to the U.S. economy has been overshadowed by other issues, like the Second Iraq War. And those issues are confounding and frustrating the American public. The next chapter will discuss how America's frustrations in Iraq, along with fears about peak oil and other matters, are contributing to the emotional appeal of energy independence.

2

THE EMOTIONAL
APPEAL OF ENERGY
INDEPENDENCE

In early 2006, George W. Bush famously declared that America was "addicted to oil." That surely is true. America is addicted to oil. And that habit is big and costly. But pardon me for asking an impertinent— but critical—question: So what?

Every other country on the planet is addicted to oil, too. Other developed countries—namely, Japan, Germany, and France—import nearly all of their oil and they have been doing so for many years.[1] China's oil import needs are growing faster than those of any other country. And yet the Chinese don't have a single soldier on the ground in the Persian Gulf. They are not panicking, nor are their leaders yammering about energy independence. The countries of the world, the *people* of the world, are addicted to oil because it is a remarkably flexible substance. It's compact, contains loads of heat energy, is easily transported, and can be used for a myriad of tasks, from transportation, to making plastics, to heating, to electricity production.

America's access to plentiful supplies of cheap motor fuel has contributed directly to its affluence. Energy consumption creates wealth. It is axiomatic: As energy use rises, people get richer. It's true always, everywhere. It is no accident that the countries with the highest per

FIGURE 2 Energy Use and Prosperity

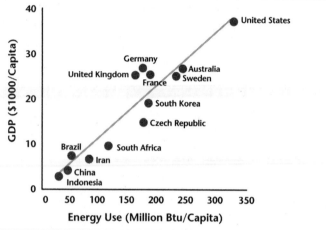

Sources: CIA Factbook, BP.

Image is courtesy of Huber and Mills, *The Bottomless Well*, and www.digitalpower group.com at http://www.digitalpowergroup.com/TBW/downloads/figure82.pdf.

capita incomes are also the ones with the highest rates of energy consumption. Nor is it accidental that the countries with the lowest levels of infant mortality tend to be the ones that have the highest rates of energy consumption. The higher living standards that come with greater energy use help foster education and slow population growth. Indeed, energy is the essential ingredient in any economy.

And yet, there is a restlessness—perhaps the best word is *guilt*— among many Americans about the amount of energy they consume. That idea was evoked by presidential hopeful Barack Obama, who, in a February 2007 speech, said America must break free of the "tyranny of oil."[2]

Obama's comment is indicative of a deep schizophrenia about energy in the U.S.

Americans love their cars, their motorcycles, their boats, and their lawnmowers, and by extension, they also love the gasoline that powers them. During the 1980s, General Motors had an advertising slogan that reflected this affection: "It's not just your car, it's your freedom." And of course, that freedom—that independence—is predicated on the availability of gasoline. (Each American uses, on average, about three gallons of oil per day.[3]) Americans not only love their cars, they love their car companies. In 2007, the giant Japanese automaker Toyota Motor ranked third in *Fortune* magazine's rankings of America's most admired companies.[4] In *Fortune*'s 2006 ranking of the most admired global companies, two automakers were in the top 15: Toyota (2nd) and BMW (13th).[5]

America may be a NASCAR Nation united in its love of the Daytona 500, the late Dale Earnhardt, and fast race cars plastered with gaudy advertisements; the U.S. may have more automobiles and more miles of paved roads than any other country on the planet; and millions of middle-aged American men may hope that their shiny new Harley Davidson motorcycles will return them to their youth; but none of these factors prevent millions of Americans from hating the oil companies with a deep and abiding passion.[6]

In an August 2006 Gallup poll, researchers found that just 15 percent of Americans had a positive view of the oil and gas industry, while 77 percent had a negative image. Out of 25 sectors that Gallup asked about, the oil and gas industry ranked dead last. Even the federal government ranked ahead (but just barely) of the oil and gas industry in the collective opinion of the general public.[7]

The American public's split personality regarding their automobiles and the gasoline they need to drive them explains some of the appeal offered by the concept of energy independence. That tormented relationship with oil and gasoline has melded with four other sticky issues in the modern American psyche. And every one of those issues is so big and intractable that together they further exacerbate the average American's anxiety about energy in general and oil in particular.

The four issues:

- Iraq
- Infrastructure attacks and Osama bin Laden
- Peak oil
- Climate change

All of those issues cause anxiety, and voters are looking for something—anything—that gives them reason for optimism. The idea of energy independence appears to do just that.

In October 2006, Democratic Party adviser James Carville sent out a memo to friends of the Democracy Corps, the political strategy group that he started with several other Democratic political consultants. Predicting a big win for the Democrats in the November 2006 elections, Carville advised that "Democrats need to talk about the change they will bring, starting with major efforts to achieve energy independence."[8] Carville, who gained fame by helping Bill Clinton win the White House in 1992, went on to discuss some polling data, which came from a group of voters who tended toward the Republican end of the spectrum. The key finding: 42 percent of those polled said that the most important national security priority of the U.S. was "reducing dependence on foreign oil." The second most important security issue, according to 26 percent of the respondents, was "combating terrorism." As Democracy Corps cofounder Stan Greenberg explained it, those two issues were also closely bound up with the Second Iraq War. "When we lay out different plans for how to deal with Iraq, any plan that also includes energy independence tops any other plan that doesn't," explained Greenberg."[9] The memo concluded with a clear message for Democrats:

> Our research shows that a candidate that says he or she will go to Washington and change things there, and will work together with both parties to do major things to move the country toward energy independence has a powerful impact on the vote. It is the one issue that gives people hope we can be more secure, get beyond Iraq, and also have a stronger economy that creates American jobs.[10]

The second sentence contains the essential message: Energy independence "gives people hope." And there's the appeal. Since the September 11, 2001, attacks, Americans have been inundated with bad news about the four issues cited above, and all of those issues are tied directly, or indirectly, to oil and energy.

First, Iraq.

IRAQ

There's a deep irony—and tragedy—in the fact that America is immersed in the rhetoric of energy independence at the very same time that the U.S. military is in Mesopotamia seeking control of Iraq's oil. Partisans can argue about the reasons behind George W. Bush's decision to launch an elective war against Iraq, but no matter how you slice it, the Second Iraq War is, in large measure, about controlling the flow of that country's oil.

Just for review purposes, it's worth recalling that the Bush administration's stated justification for the invasion of Iraq was that Saddam was supporting terrorism; further, that left unchecked, he would unleash his arsenal of weapons of mass destruction on the West. After the invasion, when inspectors failed to find any of those weapons of mass destruction—weapons that the CIA's then-director, George Tenet, and then-Secretary of State Colin Powell assured the world were there—Bush and his supporters changed their story, claiming that the U.S. had invaded Iraq in order to spread democracy in the Middle East. When democracy failed to materialize, the justification for the invasion turned, predictably, to oil.

For about two and a half decades, Saddam Hussein ruthlessly controlled Iraq—home of the world's third largest conventional oil reserves (behind Saudi Arabia and Iran).[11] That oil made him a key player in Middle East politics from the 1970s through 2003. It made him both a friend of the West and an enemy of the West. He was a friend of the U.S. in the 1970s and during the Iran-Iraq war of the 1980s. During that time, the U.S. did all it could to help Iraq increase its oil exports. He later became a foe, thanks to his invasion of Kuwait, which led to the First Iraq War. During the 1990s, Saddam's antagonism of the U.S. was made possible because of the vast oil wealth that he controlled. The oil-for-food scandal, during which Saddam siphoned off billions of dollars, was a result of Iraq's vast oil wealth. As energy journalist John Roberts explained in a 2003 paper for the International Research Center for Energy and Economic Development, oil allowed Saddam to buy weapons, foil the United Nations sanctions, and thumb his nose at the West.

"Without oil, Saddam was nothing," Roberts wrote; "with oil, he had
real power, even in adversity."[12]

Getting rid of Saddam was a fairly easy exercise for the U.S. military,
which in 2003 routed the Iraqi military and captured Baghdad in less
than three weeks. But the strategic focus of the U.S. invasion was always on oil. The very first objectives of the invading U.S. forces included the capture of key Iraqi oil terminals and oil fields. On March
20, 2003, U.S. Navy SEALs engaged in the first combat of the war when
they launched a surprise invasion of the Mina al-Bakr and Khor al-
Amaya oil-loading terminals in the Persian Gulf. A few hours later,
Marine Lieutenant Therral Childers was killed while fighting for control of the Rumaylah oil field in southern Iraq. He was the first U.S.
soldier to die in combat in the Second Iraq War.[13]

Oil was the objective when U.S. forces got to Baghdad on April 8.
Although the National Library of Iraq, the National Archives, and the
National Museum of Antiquities were all looted and, in some cases,
burned, the oil ministry building was barely damaged. That's because
a detachment of American soldiers—along with a half dozen assault
vehicles—was assigned to guard the ministry and its records.[14]

Controlling Iraq's oil was a critical element of the neoconservatives'
war plan. The prowar boosters in the Bush administration promised
that oil money was going to rebuild Iraq after the U.S. military took
control. As Michael Gordon and Bernard Trainor write in their history
of the invasion, *Cobra II*, "The Pentagon had promised that the re-
construction of Iraq would be 'self-financing,' and the preservation
of Iraq's oil wealth was the best-prepared and -resourced component
of Washington's postwar plan."[15]

The postwar plan looked good on paper while it was housed within
the safety of the Pentagon. It fell apart when it got to Iraq. The too-
small U.S. force was no match for Iraqi saboteurs, who began bombing
pipelines and other oil-related targets in early June 2003. And those attacks have continued nearly nonstop. The result of Iraq's crippled oil
infrastructure has been the unleashing of a carnival of corruption.
With the Iraqi government unable to provide the energy needed by the

April 11, 2004: A U.S. Marine punctures a jug of black market gasoline in al-Mahmudiyah, al-Anbar Province, with his bayonet. Early in the occupation of Iraq, the U.S. military tried to police Iraq's black market. It eventually gave up. By August 2006, fuel shortages were commonplace and black market fuel in Baghdad was selling for $3 to $5 per gallon. According to one source, some service stations in the city had two lines: one for regular customers and another for those paying bribes. Also note that the Humvee in the background of this photo is not armored. By 2007, virtually all of the Humvees in Iraq had been armored in an effort to blunt the effect of roadside bombs.

Source: Bill Gertz and Rowan Scarborough, "Inside the Ring," *Washington Times*, September 1, 2006.

Photo credit: U.S. Department of Defense, Corporal Carl A. Atherton, U.S. Marine Corps.

country's civilian population, the black market went wild. By early 2007, gasoline in Baghdad was so scarce at government-run service stations that black market operators were selling it for up to $5 per gallon. Cooking fuel and diesel fuel were also being sold on the black market. In southern Iraq, about 100,000 barrels of crude oil per day that were being shipped through a Basra pipeline were simply disappearing.[16] Revenues from the black market in crude, gasoline, and other oil products were then being used by Iraqi insurgents to fund their operations and continue targeting U.S. soldiers with roadside bombs, rocket-propelled grenades, and snipers.[17]

By late 2006, Bush himself was making the case that America's main reason for pressing ahead in Iraq was, in fact, about controlling the country's oil. During an October 2006 press conference, he said the U.S. could not "tolerate a new terrorist state in the heart of the Middle East with large oil reserves that could be used to fund its radical ambitions or used to inflict economic damage on the West."[18]

While Bush's claims are debatable, there's no question that a key problem for the U.S. military in Iraq is that it has never gained control of Iraq's oil. As energy economist and professor of economics at Ohio Northern University A. F. Alhajji has rightly declared, "Whoever controls Iraq's oil, controls Iraq."[19] By mid-2007, it was obvious that no one group was in control of Iraq's oil. And that lack of control was a key reason for Iraq's descent into chaos. Without a functioning oil sector and the employment and stability that it brings, Iraq was unable to provide enough fuel (gasoline, diesel fuel, liquefied petroleum gas) to keep its economy functioning. In addition to fuel shortages, electricity was scarce or nonexistent.

While the oil industry in Iraq languishes, the country has devolved into chaos as rival Sunnis and Shia launch repeated bloody attacks on each other. The fighting and devastation have led to the biggest refugee crisis in the Middle East since the Palestinian exodus from Israel in 1948. By one estimate, some 1.7 million Iraqis have been displaced inside the country. Hundreds of thousands of Iraqis have left the country, including up to 40 percent of its middle class.[20]

The refugees may be the lucky ones. According to Iraq Body Count, by July 2007, as many as 73,500 Iraqi civilians had been killed during the first four years of the war.[21] Another study—which was immediately attacked by the Bush administration and its allies as unreliable—was published by *The Lancet,* Britain's premier medical journal, in October 2006. It estimated that more than 650,000 Iraqis had died since the start of the war. An Iraqi physician who helped collect data for the *Lancet* study later said that "people outside Iraq do not realize the real disaster we are suffering."[22]

There have been heavy losses among the American military. By October 2007, the war in Iraq had killed more than 3,800 American soldiers and left tens of thousands of others wounded.[23] On the battlefield, the vaunted U.S. military—despite its vast advantages in firepower, mobility, and technology—was slowly being picked apart by insurgents, usually armed with AK-47s, rocket-propelled grenades, and roadside bombs.

There have been huge financial costs. An early 2006 study by Nobel Prize–winning economist Joseph Stiglitz and Harvard lecturer Linda Bilmes estimated that the entire bill for the war—including health care for wounded veterans and opportunity costs—may ultimately reach a staggering *$2 trillion.*[24] That's $2,000,000,000,000. For comparison, that's five times as much as Saudi Arabia's gross domestic product in 2006.[25]

By May 2007, according to the Congressional Research Service, the long war on terror had already cost U.S. taxpayers some $610 billion, and bills were being added at the rate of about $12 billion per month, a 38 percent increase over the costs being incurred in 2006, when the war effort was costing $8.7 billion per month.[26] The Iraq war alone was costing nearly $3,000 *per second.*[27] And the cost of keeping one U.S. soldier on the ground in Iraq had reached a staggering $390,000 per year.[28]

But the huge costs of the war will not stop when the U.S. finally pulls out of Iraq. Rebuilding the U.S. military and replacing lost equipment will likely cost tens (or hundreds) of billions of dollars. The U.S. has lost prestige in the international community. And the long-term damage has yet to be assessed. One military analyst, Donald Vandergriff, a former officer in the U.S. Army who has written several books and is an expert on military training, calls the Iraq war the "greatest single strategic error in American history." And ever since the war effort began to falter in late 2003 and early 2004, the American people have been looking for something to help them make sense of America's entanglements in the Middle East. Amid the avalanche of terrible news out of Iraq, Americans want to be hopeful about something. After years of the prevarications and fabrications put forward by the Bush administration, they want an idea in which they can place their faith. And if that concept can somehow encompass Iraq, then that's all to the good.

Energy independence appears to be that concept—the vessel into which Americans can pour their frustrations and their hopes for the future. As Carville explained to his clients, energy independence is an issue that voters believe can help the country "get beyond Iraq."

OIL INFRASTRUCTURE
ATTACKS AND BIN LADEN

While the U.S. has tried to gain control of Iraq's oil, the world's most notorious terrorist, Osama bin Laden, has repeatedly said that his hatred of the West stems, in part, from his disgust over America's consumption of Arab oil. That anger over oil has led him to call for attacks on pipelines, refineries, and other oil-related infrastructure.

And those attacks on oil infrastructure have led some of America's energy isolationists to insist that the best remedy to the problem is to use less foreign oil. In 2004, one of those leading energy isolationists, Gal Luft, a former lieutenant colonel in the Israel Defense Forces and now the head of the Institute for the Analysis of Global Security, wrote that "oil terrorism is now emerging as one of the biggest threats to global economy [sic]." He went on, saying that pipelines, tankers, and other oil infrastructure are "soft targets which can be easily sabotaged by those willing to sacrifice their lives."[29]

It is not at all clear that oil-related targets are "easily sabotaged." But it is clear that bin Laden has an oil fixation. In a March 1997 interview

March 1, 2006, 4:33 p.m.: U.S. Army private first class Mark Hexum stands guard after an attack on a pipeline near Taji, Iraq.
Photo credit: U.S. Department of Defense, Petty Officer First Class Michael Larson, U.S. Navy.

with CNN's Peter Arnett, bin Laden said that oil prices were "not realistic" because the Saudi royal family was "playing the role of a U.S. agent," and that the Saudis were "flooding the market that caused a sharp decrease in oil prices."[30] He went on to say that the U.S. wants to "occupy our countries, steal our resources."[31] The following year, in an interview with al-Jazeera, he again talked about oil, saying that the U.S., Britain, and Israel were on the side of the "global Crusader alliance" and that it was "not acceptable" that they "should attack and enter my land and holy sanctuaries, and plunder Muslims' oil."[32]

In 2002, in a letter that was posted on the Islamist Web site, al-Qala'h, bin Laden listed the reasons why he and his followers are attacking the U.S. Among them: "You steal our wealth and oil at paltry prices. . . . This theft is indeed the biggest theft ever witnessed by mankind in the history of the world."[33]

In December 2004, as oil prices began creeping upward, bin Laden said, "The biggest reasons for our enemies' control over our lands is to steal our oil, so give everything you can to stop the greatest theft of oil in history." And he implored his followers to attack energy infrastructure: "So keep on struggling, do not make it easy for them, and focus your operations on it, especially in Iraq and the Gulf, for that will be the death of them."[34]

Al-Qaeda operatives have heeded bin Laden's calls. In April 2004, three bomb-laden boats attacked both of Iraq's oil terminals in the Persian Gulf. The first boat headed for the terminal known as Khor al-Amaya and was intercepted by a U.S. Navy speedboat. As the sailors on the speedboat approached the insurgents' boat, the suicide bombers blew up their boat, killing two American sailors and wounding four others. Twenty minutes later, and a few miles south, two more boats piloted by suicide bombers attacked Iraq's main oil terminal, Mina al-Bakr. They, too, blew up their boats before they reached their target.

In a statement released right after the bombing, al-Qaeda leader Abu Musab al-Zarqawi reminded readers of al-Qaeda's October 2000 suicide bombing attack on the USS *Cole* while the ship was being refueled in Aden, Yemen. That attack killed 17 sailors and injured 39.[35] Al-Zarqawi claimed his loyalists "have repeated this attack in a new garb and with stubborn determination by striking vital economic links of the infidel and atheist states."

On February 24, 2006, Saudi security forces thwarted an attack on the world's single most important piece of energy infrastructure: Abqaiq. Located near Dhahran in Saudi Arabia's Eastern Province, the Abqaiq facilities process about 7 million barrels of crude oil per day—about two-thirds of the country's daily output and around 8 percent of the world's daily consumption. The attackers attempted to drive two explosive-laden vehicles into the Abqaiq compound, but they were stopped when security forces fired on them. The vehicles exploded, killing all of the attackers and three security guards.[36] The attackers were later identified as being part of al-Qaeda.[37]

That attack on Abqaiq came just two months after al-Qaeda leader Ayman al-Zawahiri exhorted his jihadists to attack oil infrastructure. In a December 7, 2005, videotape, al-Zawahiri told his men to focus "attacks on Muslims' stolen oil."[38]

In April 2007, the Saudis announced that they had arrested 172 men who were allegedly planning to attack various oil installations in the kingdom. The group, reportedly linked to al-Qaeda, had received training in Somalia, Afghanistan, and Iraq. The arrests of the suspects came just six months after Saudi law enforcement officials rounded up 136 people, all of whom were charged with planning to commit similar attacks.[39]

Oil infrastructure attacks have become commonplace in Iraq. By March 2007, some 400 attacks on Iraq's pipelines and other infrastructure had been documented.[40] And that estimate is undoubtedly far lower than the actual numbers.

The attacks on oil installations are not limited to the Persian Gulf. In fact, it's safe to say that most attacks on oil installations that occur around the world have nothing to do with al-Qaeda. For years, Nigeria's oil output has been hampered by corruption and rebel activity in the Niger Delta, where insurgents and thieves have repeatedly targeted oil infrastructure.

In Colombia, insurgent groups have a favorite target in their war on the federal government: the Caño Limón-to-Coveñas pipeline. The 100,000 barrel-per-day, 480-mile-long pipeline transports crude oil from the Caño Limón oil field (discovered in 1983 by Los Angeles–based Occidental Petroleum) in the northeastern state of Arauca to the coast. Between 1996 and mid-2005, there were some 635 attacks on the pipeline.[41] Despite the efforts of the U.S. military to protect the pipeline, the bombings and killings along the pipeline continued into 2006. For instance, in July 2006, the line was bombed. When oil field workers tried to repair the damaged pipeline, they were hit by another bombing, which killed two workers and two soldiers.[42] The bombings have been blamed on a rebel group called the Revolutionary Armed Forces of Colombia, known as FARC.

While all of these attacks are important, they are nothing new. Attacks on oil infrastructure have been occurring for decades.

In the late 1930s, "Arab terrorist bands" were regularly attacking the pipeline that carried Iraqi crude from Mosul to the port at Haifa. Between 1936, when the line opened, and mid-1938, the pipeline had been attacked some 120 times, and according to the *Chicago Daily Tribune*, "Special police patrols along the line have proved insufficient to cope with the saboteurs."[43]

During World War II, oil facilities were constantly being attacked by both sides. Early in the war, German submarines had great success targeting American oil tankers. In early 1942, as explained by Robert Goralski and Russell W. Freeburg in their epic book *Oil and War*, German U-boats "moved from Cape Cod to Cape Hatteras with impunity, sinking forty-four unarmed and unescorted merchantmen. More than 70 percent of the tonnage lost was in tankers."[44]

In 1956, during the Suez crisis, Arab nationalists blew up pumping stations in Syria that pumped Iraqi crude to the Lebanese port of Tripoli.[45] In 1967, Israeli soldiers captured the famous piece of Syrian real estate called the Golan Heights. In doing so, they took control of a pumping station on the Tapline, the 1,040-mile-long pipeline that carried crude from Saudi Arabia's Eastern Province to the Lebanese port at Sidon.[46] Two years later, the Popular Front for the Liberation of Palestine bombed the Tapline near the Golan Heights.[47]

During the Vietnam War, both sides were constantly attacking oil-related targets. In 1965, a force consisting of some 200 Viet Cong fighters attacked an air base near Da Nang. The Viet Cong then used mortars to destroy some 2 million gallons of jet fuel stored at the base.[48] In mid-1966, Lyndon Johnson's national security adviser, Walter Rostow, sent him a memo that said a "systematic attack on oil would almost certainly kill fewer North Vietnamese civilians than generalized harassment."[49]

In early 1973, as tensions between the Israelis and the Arab states grew, the Saudi oil minister, Ahmed Zaki Yamani, warned the U.S. that the Saudis' oil fields were highly vulnerable to acts of sabotage by terrorists. The *New York Times* reported that Yamani told the Nixon administration that "the risk of terrorism could be diminished by progress toward a settlement" of the Israeli-Palestinian issue.[50]

These days, while it's clear that bin Laden and his jihadis are targeting oil infrastructure, the truth is that they have not been hugely successful. Given the vast amount of oil that is transported around the globe on a daily basis, the surprising thing about oil-related terrorism is that it happens so rarely.

Oil producers around the world must accept the fact that oil infrastructure has become one of the main battlegrounds in the fight against al-Qaeda. The result is the ongoing militarization of oil infrastructure all over the world, from the Shatt al-Arab in the uppermost regions of the Persian Gulf to the Houston Ship Channel. As long as oil is produced, the infrastructure needed to support its production will always be targeted. And while security efforts can help control those attacks, there is no way to ensure that all oil infrastructure will remain free from assault. Oil-related terrorism was a problem in Iraq in 1938. It's still a problem now. And energy independence won't do anything to stop those attacks.

The point is clear: Oil-related terror attacks will continue. And those attacks will likely lead to temporary price increases in the global price of oil. But even if the U.S. were somehow free of imported oil, American consumers would still be buying oil at the global price and would therefore not be immune from the price spikes caused by infrastructure attacks. Why? Well, because the price of oil is set globally. American oil traders are not going to voluntarily sell their domestically produced crude in the U.S. if they can get a substantially higher price in, say, London or Rotterdam. Traders always seek the best price for their commodities. Thus, the U.S. *cannot* isolate itself from the rest of the global oil market.

PEAK OIL

Since September 11, 2001, Americans have seen a deluge of books, documentaries, and magazine articles about peak oil—the concept that the world's oil producers have reached the limit of their ability to produce ever larger amounts of crude. Of those many reports, five deserve to be mentioned.

In 2001, just three weeks after September 11, Kenneth Deffeyes, a geologist from Princeton University, published *Hubbert's Peak: The Impending World Oil Shortage.* Deffeyes wrote his book by building on the work of M. King Hubbert, a geophysicist who worked for Shell Oil in Houston. Hubbert studied oil production trends in America, and in 1956, using mathematical models, he predicted that U.S. oil production would peak in about 1969. He missed it by a year. Production peaked in 1970.

Deffeyes used Hubbert's models to explain that a peak in global oil production was looming, and that a catastrophe would soon follow. "An unprecedented crisis is just over the horizon," Deffeyes wrote. "There will be chaos in the oil industry, in governments, and in national economies. Even if governments and industries were to recognize the problems, it is too late to reverse the trend. Oil production is going to shrink."[51] After his book was published, Deffeyes got enormous amounts of attention from the media. He even was so bold as to predict the actual date when the world would hit peak oil production: Thanksgiving Day, 2005.[52]

Another book that falls roughly into this category is *Blood and Oil: The Dangers and Consequences of America's Growing Petroleum Dependency,* written by Michael Klare, a professor at Hampshire College in Massachusetts. Published in 2004, Klare's book does not address peak oil directly. Instead, it focuses on America's militarization of the Middle East and claims that America's thirst for oil and its reliance on foreign oil are "increasing the risk of turmoil and conflict" in the world's oil-producing countries, locales that, he points out, are, in many cases, already unstable.[53] Klare's book offers an excellent primer on the ongoing militarization of the oil sector, from the Middle East to Colombia. Alas, he concludes that the U.S. needs "autonomy" when it comes to energy, and he despairs that mere "lip service is being paid to the need for energy independence."[54]

The End of Oil: On the Edge of a Perilous New World, published in 2004 by journalist Paul Roberts, warned that the world was running out of oil and that the West's reliance on Middle Eastern oil has "helped foster a perpetual state of political instability, ethnic conflict, and virulent nationalism in that oil-rich region."[55] Roberts's book warns that our current energy system is failing. After more than 300 pages of a mostly clear-eyed survey of the global energy business, he concludes that every year that goes by is "another year in which our unstable energy economy moves so much closer to the point of no return."[56] While there is much to recommend in Roberts's book, he fails to provide readers with an understanding of the rapid globalization of the energy business. Instead, he sticks with a U.S.-centric approach, and in doing so, he implies that America holds the keys to the global energy future. That's no longer true.

Another notable book that raised anxiety about future oil production was *Twilight in the Desert: The Coming Saudi Oil Shock and the World Economy,* written by Matthew Simmons, the founder and chairman of the Houston-based investment banking firm Simmons & Co. International. Published in 2005, the book gained particular attention because Simmons is a Republican and was an adviser to George W. Bush on energy issues. Simmons's book claims that Saudi Arabia has been obscuring the problems in its oil fields and that the kingdom will not be able to significantly boost its oil output. Simmons looks at Ghawar and a number of other major oil fields in the kingdom and claims that output from those fields is declining rapidly. Simmons concludes that after those big fields go into decline, "There is nothing remotely resembling them to take their place."[57] The book sold well and garnered so much attention that the Saudi embassy in Washington took the unprecedented step of holding a public meeting to counter Simmons's claims.

The most ominous of the we're-running-out-of-energy books is James Howard Kunstler's 2005 tome, *The Long Emergency: Surviving the Converging Catastrophes of the Twenty-First Century,* which predicts that the end of inexpensive oil will lead to a total meltdown of American (and global) society. The jacket flap of the book's hardcover edition breathlessly claims that the U.S. is teetering on the precipice of disaster and that "no combination of alternative fuels will permit us to run things the way we are running them, or even a substantial fraction of them. We will have to downscale every activity of everyday life, from farming, to schooling, to retail trade. . . . Epidemic disease and faltering agriculture will synergize with energy scarcities to send nations reeling."[58]

Kunstler's book is an unrelentingly pessimistic screed that's replete with warnings that Americans must abandon virtually every familiar aspect of their current lives. He claims that the rise in energy prices will mean that the states of the Great Plains and Rocky Mountains are "destined to become the most depopulated, unproductive, and desolate places in the nation."[59] After 307 pages of dire predictions, doom, and gloom, Kunstler was so depressed that he couldn't be bothered to provide readers with a single source note, a bibliography, or even an index. Worse still, Kunstler, who presents himself as an expert on energy, misspelled the name of the massive Saudi oil complex, Abqaiq. For some reason, Kunstler calls it "Alqaiq."[60]

Several other books, along with a panoply of articles in magazines and newspapers, including the *New York Times* and *Los Angeles Times,* as well as numerous Web sites like peakoil.com, theoildrum.com, and hubbertpeak.com, have spotlighted the issue of peak oil.[61] The worries about peak oil gained further traction in July 2007 when the National Petroleum Council, an advisory group to the Department of Energy, issued a report concluding that "it's a hard truth that the global supply of oil and natural gas from the conventional sources relied upon historically is unlikely to meet projected 50 to 60 percent growth in demand over the next 25 years."[62] That same month, the International Energy Agency predicted that crude oil supplies will tighten in future years, with a "supply crunch" beginning after about 2010 as the world's major oil producers run out of spare production capacity.[63]

While all of these predictions are sobering, it's important to keep in mind that the idea of declining oil production—and the dire consequences that will follow that decline—is hardly new.

For decades, scientists, politicians, and energy analysts have been predicting that the world will run out of oil, and they generally predict that the decline will begin within a decade or so of the date of their prediction. In 1914, a U.S. government agency, the Bureau of Mines, predicted that world oil supplies would be depleted within 10 years. In 1939, the U.S. Department of the Interior looked at the world's oil reserves and predicted that global oil supplies would be fully depleted in 13 years.[64] In 1946, the U.S. State Department predicted that America would be facing an oil shortage in 20 years and that it would have no choice but to rely on increased oil imports from the Middle East.[65] In 1951, the Interior Department revised its earlier prediction and said that the oil on the planet would be depleted within another 13 years.[66] In 1972, the Club of Rome published *The Limits to Growth,* which predicted that the world would be out of oil by 1992 and out of natural gas by 1993.[67] In 1974, biologist Paul Ehrlich, who gained fame with his 1968 book *The Population Bomb,* predicted that "within the next quarter of a century mankind will be looking elsewhere than in oil wells for its main source of energy."[68] In the 1980s, Colin Campbell, one of the most vocal of the peak oil theorists, predicted that global oil production would peak in 1989.

Suffice it to say that none of these dire predictions came true. In fact, while Campbell predicted a global peak in 1989, the reality was that, between 1989 and 2005, global production increased by about 23 percent.[69]

None of this is to suggest that the idea of peak oil is a farce or that global oil production will not peak at some point in the future.[70] The quantity of oil on the planet is finite. Further, there is no denying that the number of major new conventional oil discoveries is dwindling and supplies are tightening. At some point, the amount of oil that can be produced from the earth will hit a peak and then begin to decline. Exactly when that will happen no one can say for certain. And that lack of certainty adds anxiety to the world's biggest energy consumers.

Other factors are contributing to that anxiety. For instance, the costs associated with finding and producing oil are rising sharply, particularly for U.S.-based oil companies. In early 2007, analysts from John S. Herold Inc., the highly respected oil research firm, estimated that between 2004 and 2006, the costs of developing new oil reserves for America's biggest energy companies nearly doubled.[71]

Another factor: lack of personnel. The energy industry in general, and the American oil and gas industry in particular, is desperately short of the technical staff it needs. Many companies are eager to develop various prospects, but their main constraint is their inability to find qualified engineers, geologists, landmen, roughnecks, and other people to work on their projects. In January 2007, the Interstate Oil and Gas Compact Commission released a report that underscored the problems facing an industry with a rapidly aging workforce. Of all the people working in the oil and gas sector, "half are now between the ages of 50 and 60, while only 15 percent are in their early 20s to mid-30s. The average age in the industry is 48, with some major and super major companies reporting an average age in the mid-50s."[72] And while many universities are ramping up their engineering programs to provide more professionals, filling the gap between demand and supply could take years, or even decades.

Another problem: a dire shortage of welders. Whether the location is offshore platforms in the Gulf of Mexico, equipment fabricators, or tank builders, the entire energy industry needs more welders. In mid-2007, Michael Harter, the chairman of the Tulsa Welding School, the largest welding school in the U.S., told me that there are "3 to 10 job opportunities" for every welder who graduates from his school.[73]

The backdrop for all of these challenges is the fact that the industry has little spare production capacity. For watchers of global business, that fact should not be a surprise. The oil industry has simply adopted some of the strategies of other major industries that use just-in-time delivery systems. In every industry, spare capacity of any type (manufacturing, power generation, consumer goods) is expensive. In order to stay competitive, companies are providing only as much production capacity as is needed by their customers at any given time. They meet the demands of their customers and they do it just in time. That's what the global oil business has done. Maintaining extra oil wells or platforms that are not producing to meet immediate oil demand is just too expensive.

But even as all of these factors are restraining new production, advances in technology and rising commodity prices are spurring the development of oil deposits that just a few years ago were viewed as mere fantasy. And that reflects a truism about natural resource development: As prices for a given commodity rise, more people will work to figure out how to find and produce more of it. That means that "difficult" oil like that located in oil shale and tar sands, which is virtually worthless when oil sells for less than about $40 per barrel, becomes more viable when the price of oil is, say, $60 or $80 a barrel.

The innovations in the oil exploration business were clearly illustrated in September 2006 when Chevron, Devon Energy, and Norway's Statoil ASA announced a major discovery with a well called the Jack No. 2. The three companies found a huge oil field in the deepwater of the Gulf of Mexico, about 270 miles southwest of New Orleans.[74] The Jack well, drilled in 7,000 feet of water, found a major field in a geological area called the lower tertiary trend. That formation may hold up to 15 billion barrels of oil, an amount that could boost America's reserves by 50 percent. The Jack well shows the ingenuity and technical prowess of the global energy business. Drilled to a depth of more than 20,000 feet below the sea floor, the well cost more than $100 million.[75]

A few years ago, a discovery like Jack was beyond the realm of possibility. In fact, it was just six decades ago, in 1947, that the oil industry drilled its first offshore oil well—the Kermac 16—out of the sight of land.[76] Back then, when the Kermac 16 was drilled in just 20 feet of water, the 15 billion barrels of oil that might be found in the lower tertiary may as well have been located on the dark side of the moon.[77] The industry simply did not have the technical ability to tap them. With Jack, the oil industry has succeeded in drilling for oil—and finding it—in a location that required the use of submarines, robots, and a host of new technologies. And many of those same technologies are being used to accelerate the development of offshore oil exploration all over the world.

Just as important as the technology employed is the fact that the Jack discovery refutes the entire concept of energy independence.

A few years ago, offshore oil wells in the Gulf of Mexico would often be owned, drilled, and managed by just one company. Today, the energy sector has embraced globalism to an astonishing degree. Proof can be seen in London-based BP, which is now the single biggest crude oil producer in the U.S. That's right, *a foreign oil company produces more oil in America than any American company.* In 2005, BP produced about 825,000 barrels of crude per day, or about 16 percent of all U.S. crude output.[78]

Furthermore, Jack provides ample proof of the accelerating globalization of the energy business. The well was drilled by two American companies, Chevron and Devon, along with Statoil, which is based in Stavanger, Norway. Statoil is among a United Nations–sized contingent of energy companies that are investing their capital to help ensure that Americans will have enough fuel for their cars and homes. In early 2007, energy companies from Japan, Italy, Brazil, Australia, Britain, France, and Canada were actively prospecting in the Gulf of Mexico.[79] Statoil operates in 35 countries.[80] If America really wants energy independence, why not tell all of those foreign companies to leave the Gulf of Mexico? And while the U.S. is telling those foreign companies to leave the Gulf, perhaps it should ask BP to leave as well.

BP, Statoil, and all of the other energy companies on the planet are well aware of the concept of peak oil. But the fact that peak oil looms at some date in the future does not support any of the arguments for energy independence. Instead, it's just the opposite. As peak oil gets closer, the world energy market will become even more tightly integrated as energy producers and energy consumers work to ensure adequate supplies.

A final point about peak oil and energy availability: As world oil supplies grow tighter and demand continues to rise, prices will rise, and that will squeeze the demand coming from the poorer countries.

Of course, it's impossible to forecast just how high oil prices will rise after world oil production hits its peak. But higher energy prices will not necessarily mean disaster for the U.S. and other developed countries. Charley Maxwell, the longtime energy analyst at Weeden & Co., believes that global oil production will likely peak sometime over the next decade. After the peak, Maxwell, who has been active in the energy business for five decades, foresees "rationing by price."[81] That is, the wealthier countries of the world will still be able to afford oil at $100 or $150 per barrel. The poorer countries, and poorer citizens, probably will not. Further, as oil prices continue their upward trend, oil demand will eventually start to slow, and the market will eventually reach an equilibrium between supply and demand.

That may seem to be a mercenary outlook on future oil consumption, but the truth about energy supplies is that they have always been rationed by price. The average American can afford to consume the equivalent of nearly 3 gallons of oil products per day because residents of the U.S. are among the wealthiest citizens on the planet. For comparison, the average Pakistani uses just 0.08 gallons of oil per day, not because that Pakistani doesn't want to use more oil; it's that he or she can't afford to.[82]

Resources—whether they be Rolex watches or diesel fuel—have always been rationed by price. As peak oil approaches, the rationing of oil by price will likely become more pronounced. And while that may cause some disruptions both in the U.S. and elsewhere, it does not necessarily mean that there will be shortages of oil in the industrialized world. Instead, it likely means that the developing countries of the world—and the world's poor—may have to make do with less oil than they would like simply because they cannot afford to use more.

CLIMATE CHANGE

Of all the anxiety-inducing big picture issues now facing the planet, global warming may be the one causing the most angst. Al Gore's 2006 documentary, *An Inconvenient Truth,* brought the climate change debate to mainstream America. And just like peak oil, global warming appears to be an intractable problem that has no easy solutions and no obvious paths to success.

It's not a problem that can easily be observed. Nor is it possible to place the blame for the problem on one specific car, factory, power plant, or airplane. The anxiety caused by the myriad reports about global warming has become a sort of free-floating angst. Every consumer uses fossil fuel of one form or another, and thus, global warming is a problem caused by almost everyone. And yet it can't be solved by any one single person, or one nation, or even a select group of countries. As John Lanchester wrote in a piece published by the *London Review of Books* in early 2007, many people are reluctant to even think about global warming because "we're worried that if we start we will have no choice but to think about nothing else."[83]

There are plenty of slogans about global warming. For instance, Gore claims that he is living a "carbon-neutral lifestyle."[84] And he makes that claim even though his home in Tennessee used 221,000 kilowatt-hours of electricity in 2006, or about 20 times as much as the average residence in the U.S.[85] Further, Gore and the producers of his documentary claim in their film that "you can even reduce your carbon emissions to zero."[86]

Gore was the inspiration for Live Earth: The Concerts for a Climate in Crisis. On July 7, 2007, concerts were held in London, New York, Shanghai, and several other cities in order to raise awareness about the issue of global warming and to "help solve the climate crisis."[87] Attendees at the events were urged to "answer the call" by pledging to do things like changing "four light bulbs to CFLs [compact fluorescent lightbulbs] at my home" and agreeing to "shop for the most energy efficient electronics and appliances."[88]

Global warming has been blamed for some of the recent storm events in the U.S., including Hurricanes Katrina and Rita. Some climate scientists are claiming that more big storms, along with rising sea levels and major shifts in agricultural production, are likely to occur as global warming progresses. There are increasing calls for U.S. politicians to take action on global warming by enacting laws aimed at cutting the amount of carbon dioxide that America emits. But few economists have produced reliable estimates of exactly how much it will cost to slow or reduce the amount of carbon dioxide that the U.S. pumps out every year.

There is talk about the need for carbon sequestration—the process of storing large amounts of carbon dioxide in underground reservoirs, or through some biological or chemical process—but that procedure will likely be extraordinarily expensive. One study released in June 2007 estimated that doing carbon capture and storage on an average coal-fired power plant will cost $35 per ton.[89] That may not sound like much, but when you figure that carbon dioxide emissions from coal use in the U.S. in 2005 totaled 2.1 billion metric tons, that means an additional cost of $73.5 billion per year.[90] Why is carbon capture and sequestration so expensive? The answer is simple: The volumes of carbon dioxide are daunting. That 2.1 billion tons of carbon dioxide, if collected and compressed, would total some 50 million barrels per day. That's about 2.5 times as much volume as all of the oil consumed in the U.S. per day. Accommodating that enormous volume of gas would require the construction of massive pipelines and the drilling of thousands of wells to pump that gas into the earth.

While Gore and others believe that human-made carbon dioxide is causing irreparable harm to the atmosphere, there continue to be plenty of doubters who are fighting back. They point out that the earth's climate is a massively complex system that is constantly changing and that many factors—including solar activity, methane, carbon dioxide, water vapor, and urbanization—can affect global climatic conditions.

This book is not designed to determine who is right or wrong when it comes to the science of global climate change and the effect of anthropogenic carbon dioxide. Rather, it is meant to make what should be an obvious point: Regardless of a person's individual beliefs about global warming, there's simply no question that fossil fuels will continue to be the dominant source of the world's energy for decades to come. Coal will continue to provide the bulk of our electricity, and oil will continue to provide nearly all of our transportation fuels. Natural gas will become increasingly important for heating and electricity production. And those facts will be true no matter how many Live Earth concerts are held and no matter how many people pledge to change their lightbulbs or buy energy-efficient appliances.

Furthermore, while environmentalists don't like to discuss this fact, it's abundantly clear that fossil fuels will be needed to adapt to any future changes in the globe's climate. If the earth does, in fact, get markedly warmer, consumption of fossil fuels, and coal in particular, will likely increase as the need for refrigeration and air-conditioning increases. Any large-scale attempts at carbon sequestration will also likely have the effect of increasing the amount of fossil fuel consumption. Why? The reason is simple: Moving large quantities of carbon dioxide into underground caverns, or converting that gas into another form that can be stored, will take lots of energy.

Global climate change has clearly ignited much concern. And many governments around the world are eagerly trying to address the issue of carbon dioxide emissions. But the cold, hard, inconvenient truth is that trillions of dollars have been invested in the existing energy infrastructure, which provides consumers with electricity, gasoline, jet fuel, and myriad other commodities. Changing that infrastructure—nearly all of which has been built upon fossil fuels—to a system based on renewable and alternative energy will take decades.

Among the most prominent skeptics about a rapid transition away from fossil fuels is Vaclav Smil, a polymath, author, and distinguished professor of geography at the University of Manitoba. In a 2006 speech delivered at a conference sponsored by the Organization for Economic Cooperation and Development in Paris, Smil—one of the world's most authoritative writers about energy and the history of technological advances—said that energy transitions are "deliberate, protracted affairs." Today's energy technologies are "still dominated by prime movers and processes invented during the 1880s (steam turbines, internal combustion engines, thermal and hydro electricity generation) or during the 1930s (gas turbines, nuclear fission) and no techniques currently under development," Smil said, will be able to rival those technologies over the next two or three decades. He continued, "Energy transitions span generations and not, microprocessor-like, years or even months: there is no Moore's law for energy systems." (In 1965, Gordon Moore, a cofounder of Intel, noting the rapid progress in the semiconductor industry, estimated that the number of transistors on an integrated circuit was doubling every two years.) Smil continued,

"Keep this in mind when you read yet another of the casually tossed-off claims about a continent to be electrified by wind or fueled by crop-derived ethanol by 2020 or 2025."[91]

So while Gore and others argue that an immediate, massive change in the world's energy consumption patterns is essential in order to reduce carbon dioxide emissions, the truth is that we are going to be stuck with the system we have for the foreseeable future, which in this context, likely means the next three to five decades. What effects this ongoing use of fossil fuels will have on the global climate, only time will tell.

All of these issues—the Second Iraq War, peak oil, oil infrastructure, and climate change—provide the basis for Americans' worries about energy use. And this obsession gets bolstered by a number of false promises about energy independence. The next chapter discusses those false promises.

3

THE FALSE PROMISES OF ENERGY INDEPENDENCE

The emotional appeal of energy independence is undeniable. Freedom from foreign oil—and therefore, foreign entanglements—has much intrinsic appeal. But over the past few years, the political players who are promoting the concept of energy independence have created a set of false promises to bolster their campaign and give it the appearance of credibility. Each of those false promises has to be debunked.

Let the debunking begin.

FALSE PROMISE 1:
ENERGY INDEPENDENCE MEANS THE U.S.
CAN ABANDON THE PERSIAN GULF MILITARILY

In January 2007, Set America Free, an organization based in Washington, D.C., that has been a leader in the anti-foreign-oil campaign, issued a press released titled "The Hidden Cost of Oil." The release touted a study saying that the U.S. military's "fixed costs of defending Persian Gulf oil amounted to $49.1 billion annually." It went on to say that the estimate had been done before the outbreak of the Second Iraq War. Given the war, Set America Free said, it had increased its estimate for the "additional outlays that could be reasonably assigned to the protection of oil supplies." The bottom-line military cost, according to Set America Free, of safeguarding Persian Gulf oil: $137 billion per year.[1]

Implicit, but unstated, in this presentation is that if only the U.S. used less oil—or better yet, used no oil at all—then America could save that $137 billion. With no need for oil, the U.S. could abandon the Persian Gulf from a military standpoint and never worry about it again.

Those ideas are flat wrong.

While there's no question that the U.S. Defense Department wastes billions of dollars per year thanks to hypercomplex weaponry, fraud, and mismanagement, the idea that an energy independent America will be able to abandon the Middle East—and therefore save that $137 billion or whatever is the actual cost of America's presence in the region—ignores the fundamental realities of the global energy market and the strategic and economic importance of the Persian Gulf. The countries whose coastlines touch the Persian Gulf contain about 62 percent of the world's total proved conventional oil reserves and 40 percent of the world's proved gas reserves.[2]

Given the size of those reserves and their importance to the world economy, every country on the planet—and particularly the major industrial countries—has a stake in ensuring peace and relative stability in the Persian Gulf. Political upheaval means potential supply and price disruptions. Every oil-consuming country has an interest in ensuring stable supplies and stable prices. Of course, the U.S., as the world's largest energy consumer, has a vested interest in making sure that the flow of oil and natural gas out of the region is not impeded. But that vested interest is not only in the physical commodity; it's also in the economic damage that could occur around the world if that flow of energy is interrupted.

In short, the U.S.—as well as every other country on the planet—cannot afford to see a sustained shutdown of the energy flows out of the Persian Gulf and, particularly, from Saudi Arabia. Economists have predicted that if as little as 4 percent of the world's oil shipments are halted for a long period, crude prices could nearly triple. That would mean that a barrel of crude could cost $200 or more, instead of the $70-per-barrel price that prevailed throughout much of 2007.[3] That kind of a price surge could cause the major economies of the world to shudder and could possibly lead to a recession. Therefore, like it or not, an ongoing U.S. military presence in the Persian Gulf has become a necessity. And that means that the expenditures associated with the U.S. military presence in the Persian Gulf will not disappear if the U.S. quits buying oil on the world market.

Nevertheless, the issue of "energy security" is continually invoked as a rationale for increasing subsidies to the ethanol industry. In a 2004 report, the American Coalition for Ethanol said that what motorists pay for fuel at the pump "is only a small portion of the real cost of oil. It does not reflect the environmental, military, economic and other costs directly related to our dependence on imported oil."[4] While it's true that the U.S. needs access to foreign oil and gas, it is not the only country that needs those resources. And there's where the reasoning behind the arguments put forward by groups like Set America Free falls apart.

The dangers of a supply disruption can be seen by looking at Saudi Arabia, the world's most important oil producer. Americans love to hate the Saudis. Whether the issue is religious freedom, treatment of women, the fact that 15 of the 19 hijackers on September 11 were Saudis, the corruption of the Saudi royal family, or simply xenophobia, it's apparent that many Americans would like to see the Saudis return to their days of wandering the desert, eating dates, and drinking camel milk.

That is not going to happen anytime soon.

Despite its many troubles, Saudi Arabia is emerging as one of the most influential countries in the Middle East. And that influence is entirely a product of the country's vast oil wealth. The kingdom holds more than 260 billion barrels of crude oil. At current extraction rates—and assuming that no new reserves are discovered—Saudi Arabia can keep pumping for another six decades. Now producing about 11 million barrels of crude oil per day, the Saudis account for about 13 percent of total world oil output, and that enormous volume of production makes the Saudis the de facto leaders of OPEC.* While other

---

*Saudi Arabia's total output is controversial. Some sources claim the Saudis are only producing about 8.5 million barrels per day. The E.I.A. puts 2006 output at 10.7 million barrels per day, a number that includes natural gas liquids and other production. Available: http://www.eia.doe.gov/emeu/cabs/Saudia_Arabia/Oil.html.

countries, including Venezuela, have agitated for higher prices, the Saudis have generally sought to stabilize global oil prices at levels that are good for both consumers and producers. This fact was made clear by Saudi oil minister Ali al-Naimi in 2006 during a speech in Washington, when he said:

> Energy security cannot be maintained when prices are at extremes—too low or too high. Truly sustainable energy security for consumers and producers requires three conditions—price stability, supply and demand reliability, and affordability. These are the three pillars of sustainable energy security. Affordability applies to both consumers and producers. If producers are forced to sell their energy resources at a low price, they eventually cannot afford to make the capital investments required to maximize long-term capacity. On the other hand, producers undermine their own security when their resources are not affordable to consumers. The foundation of sustainable energy security is a price low enough to avoid harming consumers, yet high enough to assure adequate return on investment for producers.[5]

No matter how much various factions in the U.S. may bash them, the Saudis' preeminence in the global oil market will continue. And by aiming for a moderate price level, they are helping keep America's domestic oil industry alive. But more important than America's domestic oil producers is the health of the overall world economy—which depends heavily on the Saudis' ability to continue producing oil at prodigious rates. If the flow of oil from Saudi Arabia were to be halted due to a coup d'état that ousted the Saudi royal family, the world economy would likely go into a tailspin. Consumers all over the world would face dramatically higher prices on everything from food and clothing to transportation. American consumers would not have as much money to spend on imported goods. Factories in China would not be able to export as many products to the rest of the world.

America has been ensuring Saudi Arabia's security since World War II, when Franklin Roosevelt met with the Saudi king, Abdul Aziz, aboard the USS *Quincy*. Ever since then, the basic arrangement has stayed in place: The Saudis provide the world with a predictable stream of oil, and the U.S. ensures that the House of Saud will stay in power. Ronald Reagan probably summed it up best in 1981 when he said, "There is no way" that the U.S. would stand by and see Saudi Arabia "taken over by anyone that would shut off that oil."[6]

Of course, America's energy interests—and therefore, its economic interests—in the Persian Gulf go beyond Saudi Arabia. America has significant military investments in Kuwait, Qatar, Bahrain, the United Arab Emirates, Iraq, and Oman, and it is not going to abandon all of those multi-billion-dollar air bases, army depots, and transportation hubs just because American motorists might, at some point in the future, be burning more ethanol-flavored gasoline.

I am not suggesting that America should bear the full cost of maintaining stability in the Persian Gulf. Rather, I'm saying that America's military history in the region—and the Second Iraq War in particular—has shown that it cannot militarize its way to energy security. Over the long term, the U.S. and other countries are going to have to rely more on markets than militarism. All of the nations of the world have an interest in a secure and prosperous Middle East. And the costs of providing military security in the region should not be borne solely by the U.S.

This point was underscored by journalist Leon Hadar in his 2005 book, *Sandstorm*. Hadar wrote that it's not the U.S. that is most vulnerable to a major supply disruption from the region. Instead, it is the economies of Europe and East Asia that are in the most danger. Those countries "should start paying most of the costs of protecting their economic interests in the region," wrote Hadar. He continued, saying that "there is no reason why America should continue to subsidize the Europeans by providing them with free security protection of these energy resources." And he concluded, "The time has come to reassess the U.S. commitment to these states, including Saudi Arabia, and bring an end to the European free-riding."[7]

Even if the Europeans and the Asian countries heed Hadar's call to quit their "free-riding" and increase their military commitments to the region, the U.S. cannot pull all of its soldiers out of the Middle East now or at any time in the foreseeable future. That doesn't mean the U.S. military has to continue basing huge numbers of troops in the Persian Gulf. In fact, keeping large numbers of troops there could actually destabilize the region. (One of the reasons for Osama bin Laden's anger at the Saudi monarchy was that it invited the U.S. military to stay in the kingdom after the First Iraq War.[8])

But it's also true that the world economy is simply too dependent on the region's energy resources to let it fall into chaos. And it's foolhardy to assume that the U.S. would allow that to happen. That means that the U.S. will be keeping navy, air force, and army units in the Persian Gulf for a long time to come.

FALSE PROMISE 2:
ENERGY INDEPENDENCE WILL
REDUCE OR ELIMINATE TERRORISM

In the wake of the September 11 attacks, American motorists have been inundated with the message that they are funding the bad guys, that by driving cars—and in particular, big gas-guzzling SUVs—their petrodollars are being used, somewhere, somehow, by a wild-eyed jihadist who wants to kill them. This conflation of oil and terrorism has been a key argument behind the push for energy independence and the effort to stop all foreign oil imports. Politicians of all stripes are repeating variations of the same claim: "We are funding both sides of this war."[9]

In an April 2006 podcast, Barack Obama said America must "think about how we are going to wean ourselves off our dependence on oil. That's a national security imperative because right now we are financing both sides of the war on terrorism by sending billions of dollars to some of the most hostile countries on Earth."[10] In an October 2006 speech, former president Bill Clinton asked students at UCLA, "Aren't you tired of financing both ends of the war on terror?" And he added that students should be worried that the money sent to oil-producing nations "might be diverted to destructive purposes." [11]

A nearly identical message was delivered by Clinton's former CIA director, James Woolsey, during a March 2007 speech to the Virginia Soybean, Corn, and Grain Association. "The next time you pull into a gas station to fill your car with gas, bend down a little and take a glance in the side-door mirror," he told them. "What you will see is a contributor to terrorism against the United States."[12]

Another leader of the oil-causes-terrorism camp is Frank Gaffney, a leading neoconservative and the head of the Center for Security Policy, a right-wing think tank based in Washington, D.C. In a January 2007 e-mail sent out by Set America Free, Gaffney solemnly declared that "some of the hundreds of billions of dollars we transfer each year to various petroleum-exporting nations wind up in the hands of terrorists. This is not simply an addiction. It is a death wish."[13]

Ah, now there's the nut of the argument: Clinton claims money in the oil-rich countries "might" go to destructive purposes. Gaffney says "some" of America's petrodollars could be put in the hands of terrorists. And on those two words—*might* and *some*—hangs much of the rationale for the jihad against foreign oil. But neither Clinton nor Woolsey nor Gaffney nor any of their cohorts bother to explain exactly how much is "some." Nor do they bother to explain how the foreign oil that comes from such notoriously belligerent terrorist havens as Canada and Mexico—which are, respectively, the first and second largest suppliers of crude to the U.S. market—poses a threat to American security.[14] Nor do they bother to explain how the U.S. can be certain that if it quits buying oil from foreigners—be they Saudis, Ecuadorians, Angolans, or one of the three dozen other countries who supply the U.S. with crude oil—the "terrorists" (wherever and whoever they are) will run out of funding and therefore be unable to launch attacks against America or its allies.[15]

Those questions never get addressed. But the conflation of terrorism and oil persists. And it has become part of what former national security adviser Zbigniew Brzezinski calls a "culture of fear" that has gripped America since September 11. On March 25, 2007, Brzezinski wrote a piece for the *Washington Post* in which he said that since the attacks in 2001, there has been "almost continuous national brainwashing on the subject of terror."[16] (In 1999, sociologist Barry Glassner published a book on this same issue, *The Culture of Fear: Why Americans Are Afraid of the Wrong Things.*) That brainwashing effort could be seen in early 2007 when a service station opened for business in Omaha, Nebraska, that claimed to be selling "terror-free" gasoline. The station sported bright blue placards that read "Terror-Free Oil Initiative" and, below that, the names of each of the four flights that were hijacked on September 11, 2001.[17]

At the Omaha service station, placards atop the pumps told consumers: "Rest assured, when you buy TFO [terror-free oil] gas, you are assisting in the global war on terror." Motorists who buy their gas at the Omaha station are "assisting" in the war, the station claims, because it only sells gasoline "from companies that purchase their crude oil from non-Middle Eastern countries." (Never mind that the Omaha station gets its fuel from Sinclair Oil, which buys some of its oil on the New York Mercantile Exchange, which sells oil from all over the world.[18]) While a few Omaha motorists may feel better about their gasoline, the idea that all Middle Eastern countries are responsible for terrorism deserves a thorough examination. To wit, exactly which Middle Eastern countries are so objectionable?

Is Kuwait one of them? That's unlikely. The Kuwaitis are perhaps America's closest allies in the Persian Gulf. Kuwait provides key military bases for U.S. forces. The Kuwaitis are one of the biggest providers of motor fuel to the U.S. military. Without the Kuwaitis, the U.S. couldn't fight the war in Iraq. So it can't be Kuwait.

What about the United Arab Emirates? That's unlikely, too. Dubai and the other emirates also provide key support to the U.S. military. Jebel Ali port in Dubai—the largest man-made port in the world— services more U.S. Navy ships than any port in the world outside the United States.[19] So the UAE is off the list.

Is one of these objectionable countries Bahrain? The U.S. maintains a large naval base in Bahrain. And while the Bahrainis also supply a great deal of fuel to the U.S. military, Bahrain doesn't supply crude to the U.S. market. So not Bahrain.

Qatar? The U.S. Central Command has a major air base in Doha. And Qatar doesn't supply substantial amounts of crude to the U.S. market. So Qatar is out.

Iraq? For the last decade or so, Iraq has been one of America's biggest crude suppliers.[20] And now that the U.S. military is occupying Iraq, the U.S. wants to see Iraq rebuilt and stabilized. Thus buying Iraqi oil should help provide funds to the nascent Iraqi government. So Iraq isn't one.

Iran? The U.S. hasn't imported any crude oil from Iran since 1991. So not Iran.[21]

Thus the furor over foreign oil—and in particular, Middle Eastern oil—must be about Saudi Arabia, the world's single biggest oil producer. And the anti-Saudi campaign is being led by Woolsey. In his March 2007 speech to the Virginia farmers, he declared that the Saudis provide billions of dollars to the Wahhabi movement, a group of fundamentalist Sunnis. He went on to claim that the "level of funding the Wahhabis provide for terrorists in Iraq, Iran and other Middle East countries is twice the amount of the largest budget ever for the KGB."[22] The Saudis are also the obvious target of the Terror-Free Oil Initiative. In early 2007, the Web site for the owners of the service station in Omaha brandished a big headline that read: "Terror-Free Oil: Saudi Arabia is *NOT* our Friend!" (capitals in the original).[23]

That's a debatable point. While it's true that America's relationship with Saudi Arabia has grown far more contentious since the September 11 attacks, it's also true that the Saudis have been a moderating influence within OPEC and that their moderating influence has helped keep oil prices from going too high. In addition, the Saudis have, in several key instances, provided crucial military support to the U.S., particularly during the Vietnam War and the First Iraq War.[24]

Further, the Saudis, throughout 2006 and much of 2007, acted as a moderate influence within the Middle East, trying to broker deals to quell the animosity between the opposing factions in Palestine (Fatah and Hamas). The Saudis also played a lead role in trying to cool the hostilities between Israel and Hezbollah in the wake of the Israeli invasion of southern Lebanon in 2006. Sunni-dominated Saudi Arabia also acts as a regional counterweight to Shiite-dominated Iran.

That said, there's no denying that Saudi Arabia has long been a home of fundamentalist and radical Islamists. There were financial links between some Saudi-funded Islamic charities and al-Qaeda.[25] Those links were exposed by a task force commissioned by the Council on Foreign Relations, which issued a report in 2002 that concluded, "For years, individuals and charities based in Saudi Arabia have been the most important source of funds for al-Qaeda. And for years, Saudi officials have turned a blind eye to this problem."[26]

Since the September 11 attacks, the Saudi government has cracked down on radical Islamists inside Saudi Arabia. It has rounded up dozens of suspects and warned radical clerics to tone down their rhetoric. In *Thicker Than Oil: America's Uneasy Partnership with Saudi Arabia,* her 2006 book on the Saudi-U.S. relationship, Rachel Bronson, the director of Middle East Studies at the Council on Foreign Relations, summed up the situation when she wrote, "Although Saudi Arabia has made dramatic improvements in its counterterrorism efforts . . . the kingdom still has a long way to go in dealing with the full extent of the problem."[27]

Clearly, the Saudis are not blameless when it comes to the funding of Islamic fundamentalists. But the implication that the Saudis and their petrodollars are the root cause of global terrorism—and the corollary suggestion that if the U.S. simply stopped buying Saudi oil the U.S. could stop "funding both sides of the war"—ignores a number of fundamental facts about terrorism.

First and foremost among those facts: Terrorism does not depend on oil money. G. I. Wilson, a retired Marine Corps colonel, who has written extensively on terrorism and asymmetric warfare, and served 28 years on active duty—15 months of which were spent fighting in Iraq—says the conflation of oil and terrorism is a "contrivance."

In 1989, Wilson, along with military analyst William S. Lind and three other authors, wrote an article for the *Marine Corps Gazette* that laid out their theories for a new type of conflict that they called "fourth generation warfare." They predicted that wars would be "widely dispersed and largely undefined; the distinction between war and peace will be blurred to the vanishing point. It will be nonlinear, possibly to the point of having no definable battlefields or fronts."[28] After two decades of studying conflict, Wilson has become one of America's most knowledgeable authorities on terrorism. He told me that terrorism and "most insurgencies are low-tech in nature. Terrorists don't need oil money. For terrorists, the money flow doesn't come from oil, it comes from drugs, crime, human trafficking and the weapons trade."[29]

That point can easily be proved by looking at the situation in Iraq. Numerous press reports have detailed how al-Qaeda in Iraq and other insurgent groups are using extortion, truck hijackings, and the black market to finance their operations. In November 2006, John F. Burns and Kirk Semple of the *New York Times* reported from Baghdad, saying that the insurgency in Iraq was "self-sustaining financially, raising tens of millions of dollars a year from oil smuggling, kidnapping, counterfeiting, connivance by corrupt Islamic charities, and other crimes that the Iraqi government and its American patrons have been largely unable to prevent."[30]

Terrorist groups in other countries have operated for years without petrodollars. In Sri Lanka, the Tamil Tigers have repeatedly used terrorism in their quest for political independence. But there has been no credible link between the Tamils and the oil trade.[31] The same is true in Spain, where the Basque separatist group, ETA, has used terrorism to target journalists and others.[32] In Ireland, in the latter decades of the 20th century, the Provisional IRA frequently used car bombs to terrorize civilians. But it did not depend on petrodollars. Nor does the Taliban, one of the main targets of America's fight against terrorism in Afghanistan. Instead, in recent years, it has relied on the booming opium trade to help fund its operations.[33] Terrorism continues to be used by the Palestinians in their ongoing battle with the Israelis. But the Palestinians do not rely on oil money to fund their attacks.[34] Another case in point: the London bus bombings on July 7, 2005. No evidence has been found linking Saudi money or petrodollars to the

London bombers who killed 52 people.[35] Furthermore, Osama bin Laden, the world's most notorious terrorist, did not get his money from oil. His family fortune came from the construction business.

It is certainly true that oil can, and does, cause conflict. It's also true that oil money has led to terrorism. That can be seen from Nigeria, Colombia, and elsewhere. But the problem of terrorism must be kept in perspective—a difficult exercise given the ongoing hyperbole and anti-Arab, anti-Islamic rhetoric that has dominated the discussion in America since the September 11 attacks.

Second, terrorism existed for centuries before the discovery of oil or the creation of Saudi Arabia. Terrorism isn't an ideology, it's a tactic. Terrorism is a method of warfare that the weak use against the strong. As Louise Richardson, the author of the 2006 book *What Terrorists Want,* succinctly puts it, terrorism "simply means targeting civilians for political purposes."[36]

Third, the conflation of oil and terrorism implies a link between terrorism and Islam. That is simply not true. Practitioners of terrorism come from all religious groups, including Christians, Jews, and others.

Fourth, the label of *terrorist* depends wholly on the perspective of those doing the labeling. One man's terrorist is another man's freedom fighter. Paul Revere, John Adams, and Samuel Adams—three icons of America's push to gain independence from Britain—were all members of the Sons of Liberty, a group that frequently used violence to intimidate British loyalists. Although they didn't commit murder, they burned the homes of the loyalists—sometimes forcing the homeowners to watch. They used violence as a method of achieving a political goal. Those men are now American heroes.

Two of Israel's national heroes, Menachem Begin and Yitzhak Shamir, were, by any definition, terrorists. And yet both became prime ministers of Israel. Begin was a leader of the militant Zionist group Irgun, and he directed the bombing of the King David Hotel in Jerusalem, which contained the headquarters of the British Army. The bombing, at noon on July 22, 1946, killed 91 people and injured 45.[37] Despite Begin's direct involvement in the killing of dozens of innocent people, he became a respectable politician and was awarded a share of the 1978 Nobel Peace Prize.[38] Shamir was a leader of the Stern Gang, which regularly used terror tactics to further the Zionist cause. Shamir ordered the murder of a Swedish diplomat, Count Folke Bernadotte, even though Bernadotte had gone to Jerusalem as a United Nations–appointed peacemaker.[39] On September 17, 1948, Bernadotte was gunned down by Shamir's lieutenants as he was being driven through Jerusalem.[40]

Similar facts surround the case of another Nobel laureate, the former president of South Africa, Nelson Mandela. In the early 1960s, Mandela led the armed faction of the African National Congress, through which he coordinated a sabotage campaign against military targets and government installations in South Africa. During his nearly three decades in captivity, Mandela refused to renounce violence— even when offered freedom if he did so. One of Mandela's most famous quotes is "A freedom fighter learns the hard way that it is the oppressor who defines the nature of the struggle, and the oppressed is often left no recourse but to use methods that mirror those of the oppressor. At a certain point one can only fight fire with fire."[41]

Fifth, terrorism can occur almost anywhere and it can be committed by almost anyone. The world now faces what terrorism analyst John Robb calls "open-source warfare."[42] Another analyst has called the current fight on terrorism a struggle between "nations and networks." The gist of this new trend in war fighting is that insurgents can wage low-level violent campaigns against their foes and do so by creating virtual networks of like-minded fighters who can be located across town or on the other side of the world. They can use cheap, easily available technologies—cell phones, Google maps, e-mail, and others—that allow them to wage war and commit acts of terrorism for very little money.

The conflation of oil and terror has stilted the entire discussion about international terrorism and its importance. In June 2006, the London-based Oxford Research Group released a report concluding that since September 11, "international terrorism has been promoted in Washington, London and other Western capitals as the greatest threat facing the world at the current juncture." But the authors of the report say that "terrorism is actually a relatively minor threat when compared to other more serious global trends, and that current responses to those trends are likely to increase, rather than decrease, the risks of further terrorist attacks." The authors of the report say further that four factors are the "root causes of conflict and insecurity in today's world" and that they will be the likely drivers of future conflicts. Those factors are climate change, competition for resources, marginalization of the majority populations, and global militarization.[43]

Sixth, in mid-2007, there were five countries on the U.S. State Department's list of countries that sponsor terrorism: Cuba, Iran, North Korea, Sudan, and Syria.[44] Only three of those countries (Iran, Sudan, and Syria) export oil. And of those three, only one, Iran, is a major exporter.

Thus, despite the surge in rhetoric seeking to conflate oil consumption with global terrorism, there's little (if any) evidence that proves that reducing oil use will translate into a decrease in terrorist activity. Further, the idea that the U.S. can halt the flow of money to Saudi Arabia— or any other oil producer—is based on yet another false promise.

FALSE PROMISE 3:
A BIG PUSH FOR RENEWABLE AND ALTERNATIVE
FUELS WILL MEAN ENERGY INDEPENDENCE

Ethanol, biodiesel, and alternative fuels have the publicity. They have the subsidies. Unfortunately, the hype and the subsidies far outstrip the actual impact these fuels are going to have on America's overall energy use.

Before going further, I need to forewarn readers that this discussion contains quite a few numbers. It also contains a number of volumetric conversions and comparisons. Do not despair! All of these figures can be easily digested. Further, getting a grasp on basic energy production numbers is an essential part of understanding why biofuels and alternative fuels cannot replace fossil fuels.

Key Conversions to Keep in Mind

1 barrel = 42 gallons

1 gallon of ethanol = 0.66 gallons of gasoline

1 gallon of ethanol = 0.59 gallons of jet fuel

1 billion gallons per year = 65,231 barrels per day

Now that the caveat is out of the way, here's the first set of numbers: In 2006, the U.S. produced about 5 billion gallons of corn ethanol and about 250 million gallons of biodiesel, most of which came from soybeans.[45]

That 5 billion gallons of ethanol is a substantial amount of fuel. It's enough to fill Houston's Astrodome about 10 times.[46] But when compared to America's overall energy use, it is, by almost any measure, minuscule. Compare that 5 billion gallons of ethanol to the fuel needs of just one company, the world's largest air carrier, American Airlines. In 2005, the Dallas-based transportation giant used 3.2 billion gallons of jet fuel.[47] Jet fuel contains about 135,000 British thermal units (Btus) per gallon.[48] A gallon of ethanol contains about 80,000 Btus.[49] That means that all of the ethanol produced in the U.S. in 2006 is approximately equivalent to 2.95 billion gallons of jet fuel—or about 90 percent as much energy as is needed by American Airlines.

Now let's compare that 5 billion gallons of ethanol to the output of an oil refinery. In 2006, Motiva Enterprises, a joint venture of Saudi Aramco and Shell, announced its plans for an expansion project at its refinery in Port Arthur, Texas. The expansion will increase the refinery's output by 325,000 barrels per day.[50] When completed, the refinery will be capable of handling about 600,000 barrels of oil per day and will be one of the world's biggest refineries.

A proper comparison requires us to convert the 5 billion gallons of ethanol that was produced in 2006 into a more familiar number: barrels

---

More Key Conversions

1 bushel of corn = 2.7 gals. of ethanol = 1.8 gals. of gasoline

1 billion bushels of corn per year = 2.7 billion gals. of ethanol per year = 176,125 gals. of ethanol per day = 116,242 gals. of gasoline equivalent per day

To convert gallons per year into barrels per day, divide the number of gallons by 15,330.

---

per day. That 5 billion gallons of ethanol is the equivalent of about 215,000 barrels of oil per day. Thus, in a side-by-side comparison, using daily output figures, the Motiva expansion project at the Port Arthur refinery will likely produce about 50 percent more motor fuel per day than the *total* output of all of the ethanol plants in America during all of 2006.[51]

Here's another comparison: Between the first week of May 1996 and the first week of May 2007, America's gasoline consumption—even though prices more than doubled over that time period—jumped by 22 percent, going from 7.6 million barrels per day to 9.3 million barrels per day.[52] That increase of 1.7 million barrels per day translates into an increase in annual gasoline consumption of 26 billion gallons. That means that *the 5 billion gallons of ethanol produced in 2006 provides only about one-fifth of the increase in overall gasoline consumption (by volume) that occurred over the prior decade.*[53]

Now let's look at America's overall oil use. In early 2007, George W. Bush said he wanted to cut America's gasoline consumption by 20 percent over the next decade. As part of that plan, Bush pledged that the U.S. would be using 35 billion gallons of renewable and alternative fuels by 2017.[54] (In June 2007, the Senate passed a bill mandating the production of 36 billion gallons of ethanol per year by 2022.[55] In August 2007, the House passed a measure providing subsidies for biofuels, but it did not set a target volume.) Regardless of whether the target is 35 billion or 36 billion gallons, it's an awfully ambitious goal—a sevenfold increase in renewable and alternative fuel production in just 10 years.

But once again, that 35 billion gallons must be put into perspective. America uses about 140 billion gallons of gasoline per year.[56] That's a lot of fuel, but gasoline accounts for less than half of America's total oil use. (The U.S. uses about 21 million barrels of oil per day or about 321.9 billion gallons per year.[57])

Therefore, even if America attains Bush's goal of 35 billion gallons of alternative fuels per year, that amount will still account for only about 11 percent of America's total oil consumption, by volume. And since ethanol—which will likely account for the largest share of those alternative fuels—contains just two-thirds of the heat energy of gasoline, the actual percentage of oil displaced will be far less.[58] When accounting for that lower heat value, that 35 billion gallons would likely provide the equivalent of 1.5 million barrels of oil per day. That's about 7 percent of America's total oil consumption—hardly enough to make a major dent in oil imports, which account for about 60 percent of U.S. consumption.[59]

These numbers aren't surprising. Back in 1997, the Government Accountability Office issued a report concluding that "ethanol's potential for substituting for petroleum is so small that it is unlikely to significantly affect overall energy security."[60] Ethanol and other biofuels cannot significantly affect overall oil consumption patterns because they cannot replace key oil-based fuels. For instance, biofuels cannot match the performance requirements of jet fuel.[61] Nor can biofuels begin to meet America's growing hunger for diesel fuel.

That statement can be proven by a bit of elementary mathematics. The U.S. consumes about 43 billion gallons of diesel fuel per year.[62] Soybeans are a favored source for making biodiesel. Farmers can produce about 40 bushels of soybeans per acre, enough to make about 60 gallons of biodiesel.[63] Thus, to replace all of the diesel fuel burned in America with biodiesel, farmers would have to plant some 716 million acres in soybeans—that's an area about 1.6 times all the cropland now under cultivation in the U.S.[64] Put another way, *even if all of the 3.188 billion bushels of soybeans produced by American farmers in 2006 were converted into biodiesel, they would yield only about 4.8 billion gallons of diesel fuel per year. That's equal to about 313,100 barrels per day, or about 1.5 percent of America's total annual oil needs.*[65] (Remember, the U.S. uses about 21 million barrels of oil per day.)

Despite the staggering amounts of land needed to produce motor fuel from food crops like corn and soybeans, the production of biofuels continues to grow. For instance, between 2000 and 2010, the International Energy Agency expects U.S. ethanol production to have grown by 20 percent per year and biodiesel output to have grown by 60 percent per annum.[66] But even at that rate of growth, ethanol and biofuels cannot supplant oil.

Again, the mathematics is easy: The U.S. Department of Agriculture estimates that distillers can get 2.7 gallons of ethanol per bushel of corn.[67] In 2006, U.S. farmers produced about 10.5 billion bushels of corn.[68] Converting all of that corn into ethanol would produce about 28.3 billion gallons of ethanol. However, ethanol's lower heat content means that the actual output would be equivalent to 18.7 billion gallons of gasoline, or about 1.2 million barrels per day.[69] Thus, *even if the U.S. turned all of its corn into ethanol, it would supply less than 6 percent of America's total oil needs.*[70]

Furthermore, even if the U.S. were able to find another type of crop to augment the use of corn for ethanol production, the ability to produce large quantities of biofuels would still be constrained by the amount of arable land. In 2006, Bank Sarasin, a Swiss bank based in Basel that specializes in asset management services for private and institutional clients, released a report on biofuels and sustainability in which it put "the present limit for the environmentally and socially responsible use of biofuels at roughly 5% of current petrol and diesel consumption in the EU and US."[71] The same report cited a study by the International Energy Agency as saying that replacing just 5 percent of the world's oil with biofuels would require up to 20 percent of all the land on the planet that's now under cultivation.[72]

For the U.S., where corn remains the only practical feedstock for ethanol production, soaring prices may prove to be the ultimate limit on its use. In 2006, corn prices soared as ethanol distilleries gobbled up one-fifth of the U.S. corn production.[73] In 2008, ethanol producers may be swallowing half of America's corn production.[74] And that could force corn prices higher still. Those higher prices will ripple through the economy as the cost of everything from meat to sweeteners gets pushed higher.

Of course, those numbers could change if there's a radical breakthrough in the production of cellulosic ethanol, the fuel that can be made from things like grass, wood, and straw. But the commercial viability of cellulosic ethanol is like the tooth fairy, an entity that many people believe in but no one ever actually sees.

Even if a major breakthrough occurs, it could take decades for the cellulosic ethanol industry to reach viability. We can see that by looking at the history of the corn ethanol business. Even with large subsidies, it took 13 years before the corn industry was able to produce 1 billion gallons of ethanol per year. And it took about two and a half decades before corn ethanol production reached 5 billion gallons per year.[75] Even with a big breakthrough in enzymes or production technology, and even with huge amounts of capital investment and federal subsidies, the cellulosic ethanol industry is unlikely to make major contributions to the domestic fuel market in the foreseeable future.

In addition to biofuels, there is growing interest in alternative fuels, which primarily involve turning either natural gas or coal into liquid fuel. And while these synthetic fuels show some promise, they, too, face constraints. The gas-to-liquids (GTL) process makes sense only in locations where natural gas is abundant. That is, gas-to-liquids plants make sense in places like Iran and Qatar, but not the U.S. Converting coal to liquids could use America's abundant supplies of coal. But the coal-to-liquids (CTL) conversion process requires the construction of expensive processing plants that emit huge amounts of carbon dioxide. And while there is some talk about building a new CTL plant in the U.S. (Montana is considered a likely location), a new plant will be expensive and will take years to build, and its output will be unlikely to add much new supply.

As for other alternatives, like shale oil, they, too, are constrained by pricing and capital. So here's the punch line: Even with the continued growth that's possible from biofuels, CTL, and GTL, these fuels will not make America energy independent for a simple reason; their numbers are just too small. And they are too small *even if you take Bush's 35-billion-gallon target and quadruple it.*

At that level, the U.S. would be producing 140 billion gallons of biofuels and alternatives. That 140 billion gallons per year would equal about 9.1 million barrels per day and would represent a 26-fold increase over 2006 biofuel production levels.[76]

Again, that's a lot of fuel. The problem is that America's energy use keeps growing. By 2030, the Energy Information Administration expects that the U.S. will be consuming about 27 million barrels of liquid fuels per day.[77] Of that amount, the EIA expects 16.3 million barrels— or about 61 percent—will be imported.[78] Even if one assumes that every drop of the hypothetical 9.1 million barrels of biofuels and alternative fuels would count against imports, the U.S. would still be importing more than 7.2 million barrels of liquid fuels per day.[79]

It's worth noting that the results of this hypothetical best-case scenario—where imports are cut to 7.2 million barrels per day—has some historical significance. Back in 1976—the year after Gerald Ford promised that America would be energy independent within a decade— America was importing about 7.2 million barrels per day.[80]

Despite the hype, biofuels and alternatives cannot, *will not*, lead the U.S. to energy independence.

FALSE PROMISE 4:
ENERGY INDEPENDENCE WILL INSULATE THE U.S. IN CASE THERE'S ANOTHER ARAB OIL EMBARGO

This false promise may be the most pernicious of all because it is based on a series of persistent myths. Before we debunk those myths, here are two examples of the political rhetoric about the dangers of another oil embargo.

- In October 2006, during a speech in California, Bill Clinton asked UCLA students to "think of the instability and the impotence you feel knowing that every day we have to have a lifeline from places half a world away that could cut us off in a minute."[81]
- The very next month, November 2006, while stumping for Republican candidates in Missouri, George W. Bush declared that if the U.S. leaves Iraq and the country is controlled by the wrong groups, they could "use energy as economic blackmail." Bush implied that another embargo would be in the offing and that those same groups would be "able to pull millions of barrels of oil off the market, driving the price up to $300 or $400 per barrel."[82]

Both Clinton and Bush were echoing claims made by Richard Nixon more than three decades ago: that the Arab oil producers are ready, willing, and chomping at the bit to once again cut off the flow of oil. This rhetoric has no basis in reality. First, a close look at the Arab oil embargo of 1973 shows that the embargo didn't work. Second, another embargo would be nearly impossible for the oil-exporting countries to execute because of the complexity and interdependence of the global market. In addition, the growth in strategic petroleum reserves in various countries means that the world is much better prepared for any shortfalls in global oil output. And third, the petrostates in the Persian Gulf cannot afford another embargo.

The 1973 embargo was launched by the Arab oil producers in an effort to coerce the U.S. to change its policies toward Israel and force the Israelis to withdraw from the territories it had captured during the 1967 war.[83] That didn't happen. Instead, in the wake of the 1973 Arab-Israeli war, and the subsequent embargo, America experienced a surge in nationalism that not only increased resentment of the Arab states but also increased America's support for Israel—both militarily and economically. Thus, the embargo failed to achieve its political objective. That fact was acknowledged by Saudi oil minister Sheik Ahmed Zaki Yamani, who said the embargo "did not imply that we could reduce imports to the United States. . . . The world is really just one market. So the embargo was more symbolic than anything else."[84]

Yamani's statement recognizes the fact that there was a lot of cheating during the embargo. Tanker cargoes destined for countries that were not embargoed were secretly diverted to the U.S. In addition, the embargo and the Arab oil producers' move to raise prices encouraged more oil and gas production in the U.S. It also spurred conservation and efficiency measures, all of which weakened the export revenues of the Arab oil producers for years after the embargo.

The 1973 embargo caused shocks to the global oil market because the world was largely unprepared for it. That lack of preparedness, combined with surging oil demand and falling excess production, magnified the impact of the embargo. Between 1968 and 1973, oil demand in the U.S. surged by about 30 percent. At the same time, the OPEC members' excess oil production capacity fell by about 700,000 barrels per day.[85]

Since the embargo, several countries have built large crude oil stocks to prevent shortages in case of political upheaval. For instance, in mid-2007, the U.S. strategic petroleum reserve contained nearly 700 million barrels of crude—that's enough oil to supply all of America's oil needs for more than a month or to cover all of America's oil imports for nearly two months.[86] In addition, the European Union, Japan, and China have built sizable petroleum reserves.[87]

Finally, it's obvious that the Arab oil-producing countries have no desire to repeat the embargo. The 1973 embargo contributed to a global recession that reduced oil consumption and hurt their own economies.

For all of these reasons, energy economist A. F. Alhajji concludes that "any future embargo will not achieve its political objectives" and that any attempt to repeat the embargo and price increases of 1973 would instead increase global prices so that the pain of the embargo would be felt by all of the countries of the world, not just the U.S. or other targeted countries.[88] In other words, the world's energy markets are now so integrated that no oil producer can selectively target one oil consumer for punishment. That means an embargo—or even the threat of one—should not be a frightening concept for the U.S. or anyone else.

FALSE PROMISE 5:
ENERGY INDEPENDENCE WILL ALLOW AMERICA TO HALT THE FLOW OF PETRODOLLARS TO ROGUE PETROSTATES

Implicit in the arguments for energy independence is the belief that if the U.S. quits buying foreign energy, then it can deny funds to selected petrostates.

This concept was articulated by the former Speaker of the U.S. House of Representatives, Newt Gingrich, in a speech he delivered to the American Enterprise Institute on September 11, 2006. In his speech, the archconservative who led the Republican revolution during the 1990s declared that the U.S. must "develop a strategic energy policy which is explicitly aimed at making the Persian Gulf and the dictatorships less wealthy and less important." He went on to express his disgust that the members of OPEC were meeting on that same day: "Think of the symbolism of the people we have enriched who are also the funders of terrorism deciding to meet on the anniversary of the day when people they funded attacked New York and Washington. We must have a strategy of moving beyond petroleum to a much more diverse and independent system of energy."[89]

The same arguments are being made by liberals.

In April 2007, U.S. Representative Ed Markey, a Massachusetts Democrat, issued a press release in which he said that his bill requiring automakers to make more fuel-efficient cars would result in substantial oil conservation. He claimed his bill would "back out every drop of oil we currently import from the Persian Gulf by 2022."[90]

Alas, Gingrich and Markey and the other energy isolationists who proffer this argument are ignoring the reality of the global marketplace.

Oil is a global commodity. Its price is set globally, not locally. Traders at
the New York Mercantile Exchange compete with bidders from Tokyo
to Tulsa, and in the process, they are constantly determining the fair
price of oil. That price becomes volatile whenever there's a hurricane,
pipeline bombing, or terrorist attack. And no matter how volatile it
gets, every barrel of oil that comes on the market is sold. Low-quality
heavy crude containing high quantities of sulfur gets sold at a discount,
while high-quality light sweet crude fetches a premium price.

The modern oil-trading system is remarkably efficient. And that efficiency prevents any buyer or seller in the marketplace from effectively designating the final consumer of crude or refined products.
This fact was best summarized by S. Fred Singer, an emeritus professor
of environmental sciences at the University of Virginia. "We can think
of the oil market in an over-simplified way as a giant bath tub into
which oil pours from many sources, where it sells at a single world
price, and from which users purchase oil without regard for its origin,"
wrote Singer in a 2003 opinion piece for the *Washington Times*. He
continued, "It is immaterial how much oil the U.S. imports from an
unstable source. It is immaterial if our imports from Saudi Arabia rise;
if they do not sell oil to us, they will sell to someone else."[91]

And that's the absolutely essential point, the critical fact that politicians like Gingrich and Markey refuse to recognize: Whatever oil the
U.S. does not buy will be bought by someone else. Oil buyers are always seeking the lowest-cost supplier. Saudi crude being loaded at
Yanbu that isn't purchased by a refinery in Corpus Christi or Houston
will instead go to a refiner in Singapore or Shanghai.

As Scott Tinker, the director of the Bureau of Economic Geology at
the University of Texas, points out, even if the U.S. did manage to decrease its oil consumption, that decrease "won't have a global impact on
major oil exporters. In fact, quite the opposite. The big oil-exporting
countries are recognizing that the future market for them is in the Mid-
and Far East, in countries like India and China. Increased demand in
those regions will overshadow decreased consumption in the U.S.,
keeping global demand above global supply."

In other words, Americans may think they are doing the right thing by burning ethanol-flavored gasoline and driving Priuses, but whatever oil the U.S. doesn't buy from the Saudis will still get sold. Every barrel of oil that goes onto the world market is purchased by someone, somewhere. Buyers cannot deny their currency to one supplier in favor of another. Every player—no matter what percentage of their oil comes from imports, whether it's 2 percent or 90 percent—is subject to price fluctuations in the marketplace. That point was summarized in a 2002 report by the Congressional Research Service, which said that "as long as prices are determined in that market, energy independence will not free the United States from oil price shocks." The report also said that attempts to achieve energy independence were only likely to raise overall energy prices for American consumers: "If an energy independent U.S. no longer participated in the world market, prices would become less volatile but would likely be higher than at their current peak."[92]

In other words, any attempt at energy independence will be enormously expensive, and it won't insulate the U.S. from fluctuating global oil prices. Nor will it mean that the Saudis, the Iranians, the Venezuelans, or any other oil-rich regime will suddenly be strapped for cash. Those oil producers will still sell every barrel of oil they produce. And yet, there is a persistent delusion that the U.S. can control the global energy market by withdrawing from it and, in doing so, deny funds to the petrostates in the Persian Gulf and force a new reform movement in the Arab and Islamic worlds.

That delusion leads to the next false promise.

FALSE PROMISE 6:
ENERGY INDEPENDENCE WILL MEAN REFORM IN THE ARAB AND ISLAMIC WORLDS

This line has been pushed hard by Thomas Friedman, a columnist for the *New York Times*. In early 2005, Friedman, who was a staunch advocate of the Second Iraq War, proclaimed that every right-thinking person should be just like him, a "geo-green"—a moniker he invented. Friedman says that geogreens want new sources of energy that will allow the U.S. to be free of foreign oil. Reducing America's demand, he argued, will lead to an oil price crash. And that crash offers the best hope America has of remaking the Islamic world. When the oil exporters are effectively bankrupted, Friedman promised, "I will give you political and economic reform from Algeria to Iran." The regimes that are foes to the U.S. "have huge population bubbles and too few jobs. They make up the gap with oil revenues. Shrink the oil revenue and they will have to open up their economies and their schools and liberate their women so that their people can compete. It is that simple."[93]

The Institute for the Analysis of Global Security, the Washington-based think tank that launched the Set America Free manifesto and has been a key proponent of the oil-causes-terrorism theory, makes the exact same claim as Friedman, saying that "America's best weapon against terrorism is to decrease its dependency on foreign oil." The group's Web site claims that:

> If the U.S. bought less oil, the global oil market would shrink and price per-barrel would decline. This would invalidate the social contract between the leaders and their people and stem the flow of resources to the religious establishment. It will likely increase popular pressure for political participation, modernity and reformed political and social institutions.[94]

But there's no evidence—*none*—to support the assertion that an oil price crash will lead to reform. Recent history shows that the petrocrats in Algeria, Iran, and elsewhere have adjusted quite easily to changes in price. Between about 1986 and 2000, oil prices generally stayed below $20 per barrel. (The two main price spikes during that time were due to U.S. military action against Iraq. Both hiccups were brief and were followed by prices falling back below the $20 mark.) By the end of 1998, prices had fallen as low as $11 per barrel. As Alan Reynolds pointed out in a May 2005 piece published in the conservative *National*

Review Online, this prolonged period of "cheap oil did nothing to promote economic or political liberty in Algeria, Iran, or anywhere else. This theory has been tested—and it failed completely."[95]

In fact, contrary to Friedman's theory, there are examples of Arab Muslim countries that are opening their societies—at the same time that they are being fed by tankerloads of petrodollars.

Kuwait, America's main proxy in the Middle East, is almost wholly dependent on oil exports. And yet, despite the rise in oil prices over the past half decade or so, Kuwait has begun reforming its governmental institutions. Of course, Kuwait is still ruled by the emir, Sheikh Sabah al-Ahmad al-Jaber al-Sabah, and the ruling al-Sabah clan. But the Kuwaitis are opening the public debate. In June 2006, Kuwait held elections for parliament. And in a radical break with the past, Kuwaiti women were allowed to vote and run for seats in the parliament. No women won seats in parliament. But the fact that an Islamic state like Kuwait has allowed suffrage is indicative of the changes under way in the emirate. And there is a widespread belief in Kuwait that it is only a matter of time before a woman is elected.

Bahrain, another American ally in the Persian Gulf, gets 70 percent of its government revenues and 60 percent of its export revenues from oil production and refining.[96] And yet, the small kingdom is a constitutional monarchy with a bicameral legislature. But having a measure of democracy hasn't necessarily quelled Sunni-Shia rivalries. The kingdom's Council of Representatives, which is elected directly by the people, has split between Sunnis and Shia. And the Muslim Brotherhood, a group that seeks to reestablish the Muslim caliphate, holds 7 of the 40 seats in the council.[97] (One of the most influential members of the Brotherhood was Sayyid Qutb, whose ideas helped shape the terrorist organization al-Qaeda.[98])

Dubai offers another example of an Islamic state that has opened its society. Like Kuwait, Dubai (which is one of the United Arab Emirates) is ruled by a monarch. But unlike Kuwait, Dubai does not have suffrage—for women or for men. Instead, power is concentrated in the hands of its ruler, Sheikh Mohammed bin Rashid al-Maktoum.

And while Dubai is ruled by one man, the emirate has blossomed, thanks to liberal policies with regard to banking and development. Today, Dubai is one of the most modern, most vibrant regions on the planet. There is tremendous economic liberty. Women hold key jobs in the government and there appear to be few limits on the emirate's future economic prospects.

Friedman and his ilk like to talk about democracy. But they should be careful what they wish for. When Arab states do hold democratic elections, they often elect Islamic parties. That can be seen in Egypt, a country that is, at least in theory, a democracy.[99]

For decades, Egypt has been ruled by one of America's closest allies in the region, Hosni Mubarak. But Mubarak, Egypt's de facto dictator, who's been in power since 1981, is no fan of democracy.[100] Public demonstrations have been banned since Mubarak took power.[101] The most powerful political group in the country is the outlawed Muslim Brotherhood. In the 2005 national parliamentary elections, the Brotherhood won one-fifth of the seats. Mubarak has since cracked down on the Islamists, and in February 2007, hundreds of members of the Muslim Brotherhood were arrested.[102] If free and fair elections were held in Egypt, it appears likely that the Muslim Brotherhood would prevail. And that could mean a new Islamic state, not democracy and pluralism.

In terms of economic freedom, Egypt ranks far below the repressive petrostates that the Institute for the Analysis of Global Security and Friedman seem to abhor. According to the Heritage Foundation's Index of Economic Freedom, Egypt ranks 127th out of some 160 countries listed in the index. Both Kuwait (57th) and Saudi Arabia (85th) rank far above Egypt.[103] (It's worth noting that Egypt has become an energy exporter over the past few years and is now exporting liquefied natural gas to the U.S. and other countries.)

That same index shows that another petrostate—or rather, a natural gas state, Trinidad and Tobago—ranks 23rd in terms of economic freedom. Norway, another energy-rich country, ranks 30th. Meanwhile, Bahrain, which has long provided military bases to the U.S. military, ranks 39th. That's higher than such Western style democracies as France (45th) and Mexico (49th).

The idea that political and economic freedom—or the converse, political and economic repression—are intimately tied to the presence (or absence) of energy cannot be proven by looking at indexes like the one assembled by the Heritage Foundation. While it's clearly true that some oil-rich countries like Angola (149th), Iran (150th), and Venezuela (144th) have little economic freedom, it's also true that other oil-rich countries like the U.S. (4th), Canada (10th), Norway, and Mexico offer substantial economic freedom.

The fundamental problem with the simplistic analysis put forward by Friedman and his neoconservative allies at the Institute for the Analysis of Global Security is that they are putting the blame in the wrong place. The critical issue in these countries isn't the question of oil revenues; it's a problem of corruption and mineral ownership. Corruption is endemic in many countries in the world, and it is particularly problematic in petrostates, where the rulers and government officials are given control of large amounts of oil revenues. Resolving the problem of governmental corruption cannot be solved by reducing the amount of money that flows into the system. Instead, it requires systemic reform. That reform includes a global effort to provide transparency in the amounts of money paid to host governments by oil, gas, and mining companies. (For more on this, look at two efforts: Publish What You Pay and the Extractive Industries Transparency Initiative, both of which are designed to make public the royalties and payments that multinational companies pay to the governments of resource-rich countries.[104]) Perhaps even more important than the corruption issue is ensuring individual ownership of mineral rights and the protection of private property.

Americans forget that the U.S. is the only country on the planet in which individuals own mineral rights. That ownership has played a key role in the development of the middle class while also restraining the growth of the government. The repression and thievery in many petrostates occur because the mineral rights are controlled by the kleptocrats, not the people. If the U.S. wants real reform in the oil-producing countries, it should be promoting property rights, human rights, and *mineral rights*. It should be advocating civil justice systems and the rule of law. Those factors are far more important determinants of the disposition of oil wealth than commodities prices. If Friedman and his cohorts are serious about advocating reform in the oil-producing countries, they should be promoting transparency and advocating efforts to fight corruption, not obsessing about prices.

Further, Friedman and the others who are advocating a collapse in the price of oil ignore the fact that such a collapse could cause widespread calamity—not the reform that they predict. The ongoing calamity in Iraq offers a prime example of just how ruinous a major social upheaval can be. Although the meltdown in Iraq was not caused by a collapse in oil prices, it's obvious that despite years of effort and billions of dollars invested, America has not been able to transform Saddam Hussein's creaky, corrupt petrostate into a model of Western liberalism. Instead, the shock therapy of invasion and prolonged war, along with the loss of thousands of lives and billions of dollars, has devastated Iraqi society and further strengthened the country's religious extremists.

It's true that reform is needed in many petrostates. But lower prices won't necessarily lead to reform. Instead, they will likely lead to calamities in several countries, hurt America's long-term interests, and increase America's reliance on foreign crude.

Those points lead to another false promise.

FALSE PROMISE 7:
ENERGY INDEPENDENCE WILL CAUSE A COLLAPSE IN GLOBAL OIL PRICES THAT WILL BENEFIT THE U.S.

The essential element in the oil-causes-terrorism argument is that belligerent petrocrats like Iran's Mahmoud Ahmadinejad and Venezuela's Hugo Chávez depend on oil revenues to fund their regimes. If their oil revenues decline, so, too, will their power. Thus, the best way to corrode their power is to bankrupt them by crushing the price of oil.

It's an interesting theory. But its advocates rarely stop to consider that a collapse in global oil prices could have many deleterious effects that would hurt America's long-term interests.

Cheaper oil will almost certainly result in higher petroleum consumption in developing countries like China and India. The Chinese government has repeatedly increased the official price of gasoline in an effort to slow that country's insatiable thirst for oil. A dramatic drop in the price of crude would reduce China's oil import bills and thereby allow greater consumption with little cost. That would allow the Chinese economy to grow even faster—growth that will further fuel China's rise as a global power. And as China's economic growth continues, it will require ever larger amounts of oil.

Lower prices would be terrible news for Iraq. And hurting Iraq's development would run counter to Friedman's own stated goals. In early 2003 Friedman wrote that "regime change is the prize" in Iraq and that "regime transformation in Iraq could make a valuable contribution to the war on terrorism." A democratic Iraq would "have a positive, transforming effect on the entire Arab world. . . . Liberating the captive peoples of the Mideast is a virtue in itself and because in today's globalized world, if you don't visit a bad neighborhood, it will visit you."[105]

Regardless of the status of democracy in Iraq, the simple truth is that Iraq has become a colonial possession of the U.S. And higher oil prices have been among the few positive things working in Iraq's favor. Those higher prices have allowed the country to amass sizable funds for its rebuilding effort, and they have helped offset the effects of Iraq's faltering oil production, which has fallen dramatically since the March 2003 invasion. Given that Iraq will—for good or ill—be America's colonial possession in the Persian Gulf for the foreseeable future, higher oil prices are far better than lower prices.

An oil price crash could be disastrous for Mexico and could increase the flight of Mexican immigrants northward. Mexico has long depended on oil revenues to fund its government. In 2006, the Mexican government got 37 percent of its revenues from cash generated by the state-owned oil company, Pemex.[106] A fall in the price of oil would devastate Pemex, which is terribly inefficient and is struggling to keep production up at its largest and most important oil field, Cantarell. That field is the world's second largest oil field (behind Saudi Arabia's Ghawar), and production from it has been falling steadily. In 2006 alone, production from the field fell by about 8 percent, and the rapid decline is expected to continue.[107]

Pemex has plenty of other problems. The company has not invested nearly enough in new exploration and production efforts. Therefore, Mexico's oil production has likely hit its peak, and the country, long an oil exporter, may soon become a net oil importer. (Mexico already imports natural gas and refined products.) Add in Pemex's lousy financial condition, and Mexico's vulnerability to an oil price crash becomes even more apparent. Between 2001 and 2006, Pemex's total debt more than doubled to some $108.4 billion. By 2006, the company's debt nearly equaled its assets, and total debt exceeded total revenues by some $13.6 billion.[108] Given those shaky financials, a sharp drop in the price of oil could send Pemex into convulsions and, with it, the government of Mexico, which has been substantially weakened thanks to the controversial 2006 presidential contest that pitted populist Andrés Manuel López Obrador against Felipe Calderón. The race was the closest in Mexico's history, with Calderón edging López Obrador by just 0.6 percent of the vote. And it resulted in weeks of protests in Mexico City.[109]

A flood of cheap crude would short-circuit the push for greater automotive fuel efficiency, both in the U.S. and elsewhere.

Motorists respond to high fuel prices. In 2006, when gasoline prices in the U.S. hit $3 per gallon, sales of more fuel-efficient cars soared. If crude (and therefore, gasoline) prices continue to fall, motorists will happily return to their Hummers, big pickups, and SUVs. And that will, once again, set up a scenario that will allow foreign automakers like Toyota, Nissan, and Honda to capture even larger shares of the domestic and global auto market.

Cheap crude would also short-circuit the push for renewable energy. We've seen this before. The surge in oil prices that occurred after the 1973 oil embargo didn't last. As prices softened, so, too, did the interest in solar power, wind power, and other technologies. The best hope for the renewable energy sector is a sustained period of high prices for fossil fuels of all types, from coal and oil to natural gas.

Low-cost oil would increase emissions of greenhouse gases. One can argue all day about what's causing global warming. But if policymakers want to decrease carbon dioxide emissions, cheap oil is the last thing they should want.

A collapse in oil prices could result in a collapse in America's domestic oil production. We've seen this movie before. In the early 1980s, Dallas and Houston were in a frenzy fueled by high-priced oil and a river of cheap money provided by crooked savings-and-loan operators. Everyone was convinced that high prices were here to stay. That illusion ended with the oil price crash of 1986, which, by the way, was largely precipitated by unrestricted production from Saudi Arabia. The crash resulted in bankruptcies from Midland to Houston. Idle drilling rigs were cut up and sold for scrap. Skilled oil-field workers left the industry for good.

A collapse in oil prices will increase America's reliance on foreign oil. Back in 1985, when America's domestic oil production was on the upswing, OPEC countries supplied 41 percent of America's imported oil. By 1990, with domestic production decimated, OPEC's share had climbed to 60 percent.[110] (By 2007, it was still about 60 percent.) If a stint of low crude prices persists, the U.S. domestic oil industry will, once again, fall on hard times. That will mean foreign producers, which generally have lower production costs, will be able to gain in market share at the expense of domestic producers.

A long period of cheap petroleum could result in instability in key countries in the Middle East. And if that instability happens, the U.S. cannot stay on the sidelines, particularly if a key ally like Saudi Arabia or Kuwait were to get embroiled in a nasty internal conflict due to an economic crisis caused by low prices.

The punch line here is obvious: Be careful what you wish for. Cheap oil could hurt America just as much as expensive oil. In fact, it might hurt more.

FALSE PROMISE 8:
ENERGY INDEPENDENCE WILL MEAN
BETTER ENERGY SECURITY FOR THE U.S.

After the hurricanes of 2005 ravaged New Orleans and other areas along the Gulf of Mexico, several refineries in the region were damaged and unable to operate. Within a few days of the storm, gasoline shortages hit several cities in the southern U.S. The shortages were, thankfully, short-lived. The reason they didn't last: imported gasoline.

Throughout the first nine months of 2005, the U.S. imported about 1 million barrels (or less) of gasoline per day. But by mid-October 2005, just six weeks after Hurricane Katrina ravaged the Gulf Coast, those imports soared to 1.5 million barrels per day, the highest level recorded up to that time by the Energy Information Administration since it began tracking those imports in 1982.[111] That gasoline, which came from refineries in Venezuela, the Netherlands, and elsewhere, was absolutely critical.[112] Without it, shortages would have surely continued.

The experience of American motorists after the hurricanes shows the wrongheadedness of the push for energy independence. This message has been put forward many times by Big Oil. But because it's Big Oil, and everyone agrees that Big Oil is bad, few people listen. Nevertheless, people like Rex Tillerson, the CEO of Exxon Mobil, keep talking about it. In an April 2006 speech, Tillerson said that the idea of energy independence offers "false hope for energy security. Only by fortifying and multiplying international energy partnerships can we protect against shocks." He added that the "truly global solution" to potential disruptions in supply is by "strengthening energy interdependence. Global commodity markets are famous for their volatility and sensitivity to world events. The market for oil is no exception. To mitigate these effects and to achieve the energy security critical to economic progress, we must diversify our sources of supply."[113]

Whether you love or hate Exxon Mobil makes no difference. Tillerson's point is demonstrably true: Energy interdependence is a fact of life. And that can be seen from the increasing amounts of imported gasoline coming into the U.S. For instance, for the first week of May 2007, America was importing about 1.1 million barrels of gasoline per day—that's about twice as much as it was importing during that same week in 2000 and about four times as much as it was importing for that time period back in 1984.[114] Those import numbers are largely the result of America's ongoing shortage of refining capacity and the continuing growth in demand for gasoline. Those factors mean that the U.S. has no choice but to buy the gasoline it needs on the global market.

The growing demand for refined oil products in the U.S. and other countries will likely be met by huge merchant refineries that are now being built in India and other locations in Asia and the Middle East. Those refineries are designed to produce tankerloads of gasoline, diesel, or jet fuel that can be shipped to the markets offering the highest prices. By 2012, India's refining capacity will nearly double to about 4.8 million barrels per day, and Indian refiners believe that over the coming years, they will be shipping more of their output to markets in Asia, North America, and Europe. In addition, the OPEC member countries are betting big on refining. They are investing a total of about $70 billion in new refining capacity, all of which is expected to come onstream by 2012 or so.[115]

These investments show that the global market for refined oil products has begun catching up with the global crude oil trade. In decades past, crude oil was the main energy commodity that was traded among the countries of the world. Refineries were located at or near the location where their products would be consumed. Economics prohibited large-scale trading and shipping of refined products over long distances. That's no longer the case. Cargoes of gasoline, jet fuel, and other products are increasingly mobile. Refiners can use large tankers to arbitrage their production and send their commodities to the markets that offer the best prices.

The former Secretary-General of the United Nations Kofi Annan once said that "arguing against globalization is like arguing against the laws of gravity." In the case of the refined-products business, arguing that the U.S. should be energy independent at the same time that the global trade in everything from gasoline to jet fuel is accelerating makes absolutely no sense from either an economic or a security standpoint.

Back in 1913, the British statesman Winston Churchill said that "safety and certainty in oil lie in variety and variety alone." That is, it's better to have many suppliers than few suppliers. Today, the world has a number of suppliers that are capable of providing the U.S. with the crude oil, gasoline, jet fuel, natural gas, and other energy commodities that it needs to fuel its economy. Rather than running from that reality, the U.S. should embrace it.

PART TWO

FROM DOMINANCE TO DEPENDENCE

American Energy History, Rhetoric, and the New Realities

4

THE GLOBAL
ENERGY POWER SHIFT

From 1859, when Colonel Drake discovered oil in Pennsylvania, through 1973, the U.S. was the dominant player in the global energy business. For much of that time, America was both the dominant producer and the dominant consumer of oil and gas on the planet.

That dominance extended into technology, finance, transportation, and refining. When it came to developing oil reserves and getting those reserves into the marketplace, the U.S. had no serious rivals. American drill bits, like those made by Hughes Tool Co., bored the holes. American companies, like Gulf Oil, or Standard Oil of New Jersey, did the seismic work, managed the production, built the pipelines, and did the refining. The drilling work was done by companies like Sedco. The drilling technology was developed by outfits like Halliburton. The bridges or dams needed to support the cities that were created by the new oil wealth were built by Halliburton's subsidiary Brown & Root or by American engineering giants like Bechtel. Texas-based law firms like Baker Botts or Vinson & Elkins handled much of the legal work. And all the while, the prolific oil fields in Texas, Oklahoma, and other states allowed the U.S. to effectively set the global price of crude.

Those days are gone.

A half century ago, American-based energy companies pumped about 45 percent of all the oil produced overseas. Today, that percentage is about 10 percent.[1] Out of the top 20 oil-producing companies on the planet, 14 are national oil companies like Saudi Aramco or the National Iranian Oil Company. Furthermore, the national oil companies now control about 77 percent of the world's proven oil reserves. The international oil companies control less than 10 percent.[2]

American energy companies are still big players in the global market, but they are no longer the dominant players. Instead of dictating terms, American energy companies and other international energy companies must now court the national oil companies that sit atop the vast majority of the world's remaining oil and gas deposits. That means that state-controlled outfits like Saudi Aramco, Russia's Gazprom, and Venezuela's Petróleos de Venezuela (PDVSA) are, in many cases, able to dictate the rules by which the major oil companies must play.

Venezuela and Russia are prime examples of how this "resource nationalism" is changing the rules of the game. In June 2007, Exxon Mobil and ConocoPhillips both decided to give up their operations in Venezuela rather than work under the conditions set by the country's populist president, Hugo Chávez. Under Chávez, the Venezuelan government has begun demanding a controlling equity stake in all of the country's oil and gas development projects, including those in the Orinoco Belt, a region that may hold more than 200 billion barrels of heavy crude.[3] Both companies had major operations in the Orinoco and both will likely face multi-billion-dollar write-downs of their assets due to the loss of their stakes in the oil-rich region. The controlling interest in the projects will now be held by PDVSA. The move is part of what Chávez is calling "21st-century socialism," and he is using the country's oil reserves to fund programs that provide more health care and other resources to Venezuela's poor.[4]

Similar nationalistic moves are under way in Russia. In June 2007, Vladimir Putin's government effectively forced BP to give up its controlling stake in the giant Kovykta gas field in Siberia, a move that will cost BP billions in potential revenue. BP sold its controlling interest in Kovykta to Gazprom for a fraction of its real value. Gazprom agreed to pay as much as $900 million for an asset that is likely worth more than $3 billion. The British company had been held hostage by Gazprom for years. The Russian company, which is controlled by Putin's cronies, refused to build any pipelines to Kovykta, thereby preventing BP from selling gas from the field.[5]

At the same time that the big oil companies are losing their negotiating strength, rising demand from China, India, and other developing countries is allowing the national oil companies to change their focus. Instead of looking first to export their products to Western consumers, they are looking east.

The purpose of Part 2 is to give a brief summary of the key events in the ongoing power shift in the global energy business and to review the U.S. government's three-decade-long uneasiness with that shift. That uneasiness is reflected in the persistence of the political rhetoric about energy independence, which has been a recurring theme in U.S. politics since October 1973, when the Arab oil embargo shook the global economy. That theme began in earnest during Richard Nixon's State of the Union address in January 1974. In the speech, Nixon declared that by 1980, the U.S. "will not be dependent on any other country for the energy we need to provide our jobs, to heat our homes, and to keep our transportation moving."[6]

The U.S. has never come close to attaining Nixon's goal. And as Part 2 will show, the rhetoric about energy independence has shifted from one guided primarily by energy policy to one that is guided almost purely by ideology, isolationism, and thinly veiled anti-Arab sentiment. And the loudest proponents of this new ideology are the same neoconservatives who pushed hard for the Bush administration's invasion of Iraq in 2003.

But before looking at OPEC, Iraq, and the neocons, readers must understand how America came to dominate the global energy industry. And that dominance was largely due to the oil pricing power wielded by the Texas Railroad Commission.

5

1972

Texas Taps Out

Long before the rise of OPEC, and years before Saudi Arabia became the key player in the global oil business, the world's most important oil cartel was based in downtown Austin, Texas.

Between the 1930s and the early 1970s, the three members of the Texas Railroad Commission were the most important people in the global oil business. They met once per month to set "allowables": the volume of oil that each operator in the state was allowed to produce from his (or her) wells that month. The allowables were set to meet current oil demand and not a barrel more. The Texas cartel operated in a straightforward manner. The three commissioners looked at oil inventories. If they were rising, they cut production. If inventories were falling, they allowed production to rise. And because the Railroad Commission controlled the flow of oil from the world's most prolific fields—the ones in Texas—the system worked. No other entity was able to control the supply of oil with the discipline and effectiveness of the commission. And by controlling the prices in the burgeoning American market, the Texas cartel effectively determined world prices, too.

By the late 1940s and 1950s, increasing amounts of oil were being discovered in Texas, Venezuela, the Persian Gulf, and elsewhere. And those discoveries led to an enormous oversupply of oil production capacity.

So the Railroad Commission simply cut the allowables for Texas producers, thereby balancing supply with demand. Even in a potentially glutted market, prices didn't fall. In fact, they rose, giving every producer even bigger profits. As one economist explained it, the system allowed the big American oil companies to "fix their own prices and make them stick." A study done in 1949 by the U.S. Senate's Small Business Committee said the Railroad Commission's system formed "a perfect pattern of monopolistic control over oil production and the distribution thereof . . . and ultimately the price paid by the public."[1]

The Railroad Commission may have been a cartel and it may have had monopolistic control, but it also brought stability to a chaotic market. Without the cartel, oil producers were constantly being whipsawed by prices, going back and forth between boom and bust, between underproduction and overproduction, as prices rose and fell in chaotic patterns. In the absence of the cartel, producers rushed to get as much oil out of the ground as they could in order to profit before the market became even more saturated with oil.

Neighbors with wells tapping into the same field would overproduce oil from their well to ensure that "their" oil wasn't pumped out by adjacent landowners. The chaos in the American oil business reached its acme in the early 1930s, shortly after prospectors discovered the giant East Texas oil field. After several years of legal wrangling at the state and federal level, the Railroad Commission was empowered to impose production limits and "unitize" fields, thereby apportioning the underground oil rights to the owners of the land above.[2]

But Texas's dominance of the industry went far beyond legal issues and oil prices. Texas oil provided a critical advantage in World War II. The Big Inch and Little Big Inch pipelines, both of which were built in record time during the war, provided huge quantities of fuel to the East Coast and became key elements of the American war effort. (That said, it's worth noting that America's domestic oil production couldn't keep pace with demand during the war. In both 1944 and 1945, at the height of World War II, the U.S. was a net crude importer. The war years are notable for another fact: The last time the U.S. was a net crude exporter was 1943.[3])

A surfeit of Texas oil prevented the Arab oil producers from using the threat of an oil embargo to pressure European countries and the U.S. during the Suez crisis in 1956 and the Six Day War in 1967.

But America's dominance of the global oil business couldn't last forever. And the end of its dominance can be traced to a specific date: March 16, 1972.[4] At the meeting on that day, the three members of the Texas Railroad Commission met and declared "a 100 percent allowable for next month." In other words, the state's oil producers could run their wells at full capacity. Without explicitly saying so, the commissioners had admitted that the Texas oil wells had reached the limits of their productive capacity. The U.S. oil business, which, for four decades, had had near-total dominance of the world market, no longer had the ability to supply extra oil to the market and therefore drive prices down. Without that ability to produce more oil than the market needed, the Railroad Commission's power as a cartel was lost.

Although few people recognized it at the time, the commissioners' move was an inevitable result of the peak in America's overall oil production. In 1970, two years before the Railroad Commission's announcement, U.S. oil production hit its all-time high of 9.6 million barrels of oil per day.[5] And ever since 1970, America's oil production has been in a gradual decline. In 2005, U.S. oil production averaged just 5.1 million barrels per day, its lowest level since 1949.[6]

The stability and price protection that the U.S. got from the Railroad Commission was only part of America's ability to control the world's oil supply. American oil companies were also protected from OPEC by federal laws that limited the amount of oil that could be imported. In 1959, Congress mandated that no more than 20 percent of America's oil supply could come from foreign producers.

America's independence from foreign oil producers meant that a new organization, the Organization of the Petroleum Exporting Countries—which was founded in the early 1960s by Saudi Arabia, Iran, Iraq, Kuwait, and Venezuela—could not control the price of oil. Nor could the OPEC members—which generally had lower production costs than American producers—gain much market share in the U.S.

One of the biggest backers of the quota on imported oil during the 1960s and 1970s was a Republican congressman from Texas named George H. W. Bush.[7] And the language he used to justify the import quotas was remarkably similar to the rhetoric being used by today's advocates of energy independence. For instance, in early 1970, Bush spoke to an oil industry group in Beaumont, Texas, telling them that he was introducing legislation that would protect them from foreign oil. Bush's legislation was designed to further reduce the amount of foreign oil that could be imported into the U.S. to 12 percent of total demand—a decrease from the 20 percent limit that was being enforced at the time. Bush told the group that imposing the quota would stimulate oil and gas drilling in Texas and make the U.S. less dependent on foreign oil. "This is particularly true now," he told them, "when instability in the Middle East severely threatens sources of our petroleum imports from that region of the world."[8]

But neither the protectionist strategies advocated by Bush nor the Railroad Commission's pricing power would last. America's increasing oil consumption and declining oil production ensured that.[9] In April 1973, the import quota on foreign oil ended. And just six months after that, America was hit by the biggest energy crisis in its history, the Arab oil embargo.

6

1973

The Embargo, Militarism, and Rhetoric

It's common wisdom in the United States that the Arab oil embargo of 1973 was responsible for the long gas lines and tremendous upheaval that hit American society in the months after the Yom Kippur War.

That wisdom is wrong.

The embargo—which began on October 17, 1973, just 19 months after the Railroad Commission went to 100 percent allowable, and lasted until March 1974—changed the global balance of power and put America on the defensive.[1] But the embargo did not cause a shortage of motor fuel or the gas lines that hit the U.S. during that time period.

Let me repeat that: The embargo did not cause motor fuel shortages in the U.S. That's because there was no shortage of crude. Energy Information Administration data show that America's crude oil imports in 1973 exceeded 1972 levels by 372 million barrels. In 1974, imports jumped again, exceeding 1973 levels by 85 million barrels.[2] The true, but sadly underreported, fact about the 1973 embargo was that the shortages and gas lines were caused by too much government intervention in the energy market.

In 1971, Nixon imposed price controls that prevented oil companies from passing on the full cost of imported crude oil to their customers. Those price controls led the big refiners to cut back on the

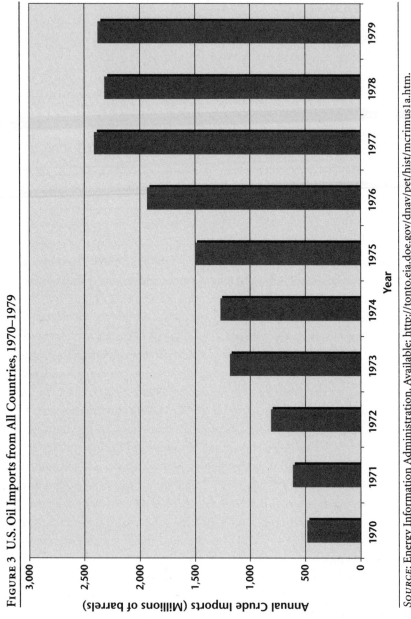

FIGURE 3 U.S. Oil Imports from All Countries, 1970–1979

SOURCE: Energy Information Administration. Available: http://tonto.eia.doe.gov/dnav/pet/hist/mcrimus1a.htm.

amount of gasoline they sold to independent service stations so that they could keep their own franchisees supplied with fuel. By mid-1973, months before the embargo began, about 1,000 service stations had closed due to the lack of available fuel. In response, about a month before the embargo began, Congress passed a law mandating that the refiners had to share their gasoline equally between independent service stations and branded service stations. It also gave each state's governor control over a small percentage of the gasoline, which further reduced the supply.

The embargo had little effect on the oil companies' ability to produce gasoline because they had stockpiled large amounts of crude in the months before the embargo began. But thanks to governmental intervention in the market, they had no incentive to refine that crude into motor fuel. In fact, they had a major disincentive, namely, that the more gasoline they made, the more money they lost. Modern-day economists looking back at the fuel shortages that occurred during the embargo place the blame squarely on governmental policies. In 2001, Donald Losman, an economics professor at the Industrial College of the Armed Forces, wrote that "most of the damage from the 1973 oil embargo emanated from our own policy blunders. Pre-embargo, poor U.S. policies made us vulnerable, and post-embargo, continued price controls and misguided regulation magnified the damage."[3]

Looking back, it's obvious that Nixon's policies exacerbated the effects of the embargo. Nevertheless, Nixon saw more bureaucracy as the cure. In mid-1973, he named former Colorado governor John A. Love as America's first "energy czar."[4] In early November 1973, Nixon said the U.S. was heading "toward the most acute shortages of energy since World War II." At the same time, he unveiled "Project Independence," saying that the U.S. should set a "national goal, in the spirit of Apollo, with the determination of the Manhattan Project, that by the end of this decade we will have developed the potential to meet our own energy needs without depending on any foreign energy sources."[5] Two months later, in his last State of the Union speech, Nixon said that the U.S. "will not be coerced" when it comes to energy.[6]

Less than seven months later, Nixon resigned rather than be impeached for his role in the Watergate scandal. But the lingering effects of the 1973 embargo and the ongoing Cold War led America's political leadership to respond to the country's energy import needs in two ways: dramatic increases in the militarization of the Middle East and regular doses of rhetoric about the need for energy independence. And despite the increase in both militarization and rhetoric, America's reliance on foreign energy has continued its upward trend.

The initial push for militarization came within a few days of the onset of the embargo. In November 1973, Secretary of State Henry Kissinger held a press conference in which he declared that if "pressures continue unreasonably and indefinitely," then the U.S. would have to consider what he called "countermeasures." Those countermeasures included a secret plan—developed by Kissinger along with Nixon's defense secretary, James Schlesinger—that called for the U.S. military to invade Saudi Arabia, Kuwait, and Abu Dhabi and seize the region's oil fields. The details of the plan, which became public in 2004, said that the three countries "would have to be securely held probably for a period of some 10 years."[7]

While the invasion never happened, the 1973 embargo marked the beginning of the shift in gravity in the global oil market away from the U.S. and toward the Persian Gulf. And rather than accept the shifting fortunes of the global energy market, Nixon and his successors have focused almost exclusively on their military options in the region.

The biggest recipient of U.S. aid in the region? Israel. In 1973, the U.S. provided the Israelis with military loans worth some $307.5 million. In 1974, the overall assistance package jumped eightfold to nearly $2.5 billion. Between 1975 and 1985, the U.S. gave Israel some $29.7 billion in aid, the vast majority of which was for military purposes.[8] In 1986, the U.S. even began providing free motor fuel to the Israeli military. That free fuel program continues to the present day.[9]

The U.S. also stepped up its weapons sales to its client in Persia, the shah of Iran. During the 1970s, the shah spent some $14 billion on

American weapons. The shah also provided troops to the sultan of Oman in an effort to repress a rebellion in Oman's Dhofar province.[10] The shah, who had been installed as Iran's ruler thanks to a 1953 coup sponsored by the Central Intelligence Agency, was almost wholly dependent on the U.S. for his political survival. And his repressive regime could not withstand the pressure of the Iranian public. In 1979, he fled after massive street protests. The resulting upheaval in Iran caused the country's oil production to fall dramatically, and the result was a spike in global oil prices.

The U.S. was quick to blame the Shiite revolutionaries in Iran for the problems in the oil market. Doing so provided justification for the continuing militarization of the Persian Gulf. But the blame for the oil price spike belongs, at least in part, to the U.S. Had it allowed the country's democratically elected leader, Prime Minister Mohammad Mossadegh, to remain in power in 1953, it is reasonable to assume that there would never have been the type of upheaval that occurred after the shah was deposed.[11] Furthermore, some of the same price controls that bollixed up the gasoline market during the 1973 embargo were still in place in 1979. And those controls reduced the ability of the oil companies to respond to the decreased supply of crude from Iran.[12]

In 1980, after the Soviets invaded Afghanistan, President Jimmy Carter, in his State of the Union speech, declared that the U.S. would not allow the Soviets to control the flow of oil out of the Persian Gulf. Carter said that the Soviets, who had moved warships close to the Straits of Hormuz, were a "grave threat" to the movement of the region's oil. He concluded with a statement that became known as the Carter Doctrine: "Let our position be absolutely clear: An attempt by any outside force to gain control of the Persian Gulf region will be regarded as an assault on the vital interests of the United States of America, and such an assault will be repelled by any means necessary, including military force."[13]

In the months that followed his speech, U.S. militarism continued. Carter negotiated a new U.S. military base in Oman. He also got new basing rights in Somalia, which gave the U.S. the ability to patrol the entrance to the Gulf of Aden and the Red Sea.[14] Carter also created a new military command within the Department of Defense called the Rapid Deployment Joint Task Force. Now known as Central Command, the task force was given the responsibility of managing all U.S. military operations in the Persian Gulf. (Today, in their briefing materials, Central Command officials note that their territory, which covers 27 predominantly Muslim countries that stretch from Sudan in the west to Kazakhstan in the east, contains 65 percent of the world's known oil reserves.)

Ronald Reagan followed Carter's lead and continued the militarization. In 1981, the Reagan administration pushed through an $8.5-billion weapons deal for Saudi Arabia that included advanced surveillance aircraft as well as aerial tankers, air-to-air missiles, and other gear.[15]

Like Carter, Reagan made it clear that U.S. policy was to protect the Persian Gulf. In 1983, Reagan signed National Security Decision Directive 114, which said that the U.S. was ready to deter any attacks on "critical oil productions and transshipment facilities in the Persian Gulf." The directive went on to state that it was American policy to "undertake whatever measures may be necessary to keep the Strait of Hormuz open to international shipping. . . . Because of the real and psychological impact of a curtailment in the flow of oil from the Persian Gulf on the international economic system, we must assure our readiness to deal promptly with actions aimed at disrupting that traffic."[16]

Reagan's concern was clearly about the price of oil. Any disruption of the flow of oil through the Strait of Hormuz would inevitably lead to a price spike. And fear of price increases remains the justification for America's militarization of the Persian Gulf.

That motivation led Reagan to cozy up to Saddam Hussein even though Saddam had instigated the Iran-Iraq war. In 1983, the Reagan administration sent a special envoy to meet with Saddam: Donald Rumsfeld. Rumsfeld's main interest in meeting with the Iraqi dictator was to help him increase his oil exports. During their chat in Baghdad, Rumsfeld and Saddam talked extensively about new pipelines that would allow Iraq to avoid the Persian Gulf, where tankers loaded with Iraqi crude could be targeted by the Iranians.[17] The Reagan administration also helped Saddam with weaponry. In October 1984, more than a month before the U.S. and Iraq restored diplomatic relations, Texas-based Bell Helicopter began talking with Iraqi officials about a big helicopter deal. Although America had pledged not to sell weapons to either Iran or Iraq, the 45 large helicopters that Bell wanted to sell the Iraqis could easily have been used for military purposes. Furthermore, American diplomats were aware that the Iraqis had used helicopters for some of their chemical attacks against the Iranians. When word of the sale first surfaced, the State Department insisted that it would not allow the sale to happen. But in September 1985, about the same time the Iraqis were attacking Iranian oil installations on Kharg Island, the U.S. approved the sale of the 20-seat 214 Super Transport helicopters, all of them built at Bell's factory in Fort Worth. The deal was worth $200 million to Bell. It also put America clearly on the side of the Iraqis in the Iran-Iraq War.[18]

In early 1987, fears about oil price increases led to more direct intervention in the Persian Gulf by the U.S. military. As the Iran-Iraq war raged on, the U.S. put 11 Kuwaiti tankers under the U.S. flag and began protecting them as they traversed the Persian Gulf.[19] The U.S. Navy even established floating firebases in the middle of the Persian Gulf as a way to help patrol the region. The barges the *Hercules* and the *Wimbrown VII* were provided by a contractor familiar with the Persian Gulf, Halliburton. Those barges were home to a group of helicopters that provided aerial support for the oil tankers traversing the Gulf.

Three years later, America's old ally, Saddam Hussein, invaded Kuwait. And once again, oil was the rationale for U.S. military intervention. At a September 11, 1990, appearance before the Senate Armed Services Committee, then-Secretary of Defense Dick Cheney said that if Saddam was allowed to stay in Kuwait, he would be in a "position to be able to dictate the future of worldwide energy policy" and that would give "him a stranglehold on our economy."[20] Cheney's rationale was repeated two months later by another top Bush official, Secretary of State James A. Baker III. On November 13, 1990, Baker held a press conference during which he—just as Reagan had done seven years earlier in National Security

Decision Directive 114—talked about potential oil supply disruptions and how that type of disruption could affect oil prices.

Baker said that the "economic lifeline of the industrial world runs from the [Persian] Gulf, and we cannot permit a dictator such as this to sit astride that economic lifeline. And to bring it down to the level of the average American citizen, let me say that means jobs. If you want to sum it up in one word, it's jobs."[21] The ultimate job-protection campaign began on January 17, 1991, when the U.S. started its aerial bombing campaign against targets in Iraq and Kuwait. The ground assault for the First Iraq War began five weeks later, on February 24.[22] After 100 hours of fighting, the U.S. military had pushed the Iraqi forces out of Kuwait, captured thousands of Iraqi soldiers, and pushed to within 150 miles of Baghdad.

For the U.S. it was a good war. It was a conventional conflict where the fighting happened between tanks and artillery. The two sides could easily be identified on maps. Soldiers on both sides wore uniforms. Fighting the Iraqi military in the open desert allowed the U.S. military to use all of its technology, firepower, and mobility. When the ground fighting ended, America was the undisputed sole superpower on the planet. And it used that status to solidify its military presence in the Persian Gulf.

In the nearly two decades that passed between the 1973 embargo and the First Iraq War, the U.S. spent uncounted billions of dollars on its militarization of the Persian Gulf and the Middle East, all in the effort to ensure that the flow of oil would be unimpeded. But even as America's militarization of the region continued, the rhetoric in American politics, particularly the rhetoric used by American presidents, continued to focus on the idea of energy independence.

In early 1975, President Gerald Ford, in his State of the Union speech, solemnly declared that the state of the union was "not good." And much of America's anxiety was due, Ford said, to worries over energy. He told the combined houses of Congress that he was considering gasoline rationing and gasoline taxes to conserve oil. "A massive program must be initiated to increase energy supply, to cut demand, and provide new standby emergency programs to achieve the independence we want by 1985," he said. To that end, Ford sought increased offshore oil drilling, expanded use of coal, expedited licensing for more nuclear plants, huge increases in the production of synthetic fuels from coal, and dozens of new oil refineries.[23] But Ford couldn't convince Congress to go along with his ambitious plans. And in 1976, Jimmy Carter unseated Ford.

Carter's energy rhetoric was much the same as that used by Ford and Nixon. In 1977, Carter declared that the U.S. had to reduce its "vulnerability to potentially devastating embargoes" and that the country had to "reduce demand through conservation." And in a preview of the current discussions about peak oil, Carter predicted that by the mid-1980s, world oil demand would outstrip supplies: "Unless profound changes are made to lower oil consumption, we now believe that early in the 1980s the world will be demanding more oil than it can produce."[24]

Carter's focus on energy reached a crescendo in early 1980, in the wake of the Iranian revolution. The ouster of the shah of Iran led to the collapse of the Iranian oil sector, and that sent prices dramatically higher. Between 1978 and 1980 crude prices more than doubled from about $9 per barrel to over $21.[25] The surge in prices led Carter to declare that "foreign oil is a clear and present danger to our Nation's security. The need has never been more urgent." He urged Congress to adopt a new energy plan that he said would include a "major conservation effort, important initiatives to develop solar power, realistic pricing based on the true value of oil, strong incentives for the production of coal and other fossil fuels in the United States, and our Nation's most massive peacetime investment in the development of synthetic fuels." Carter also proposed a limit on oil imports, which could be backed by imposing a fee on imported oil. And like Ford, he threatened to impose gasoline rationing.[26] Carter even had some 5 billion gasoline rationing coupons printed up, at a cost of $12 million, in case rationing became necessary.[27]

Perhaps the most expensive element of Carter's push for energy independence was the federally funded Synthetic Fuels Corporation. Launched in 1980, the agency provided money and loan guarantees to entrepreneurs who wanted to produce motor fuel from coal and oil shale. Congress supported the program, wrote energy analyst Vito Stagliano in his 2001 book, *A Policy of Discontent,* "even as one uneconomic project followed another, justified by the ever-elusive standard of energy security."[28] By 1992, the new federal agency was supposed to be producing 1.5 million barrels of synthetic fuel per day.[29] It never got close to that target. Instead, wrote Stagliano, it never produced "a single cost-effective barrel of fuel but managed to rack up federal debt obligations of over $2 billion."[30] (The agency was abolished in 1985.)

In addition to the push for synfuels, Carter dialed back the thermostat at the White House, donned a cardigan, and warned Americans that energy shortages were imminent. He also proposed a "new solar strategy" that called for the U.S. to get 20 percent of its electricity from solar power by the year 2000. To that end, in 1979, he had four arrays of solar panels installed on the roof of the West Wing. The panels, which cost $28,100 (about $84,000 in 2006 dollars), were used to heat water for the staff mess and other parts of the White House.[31] When Carter showed the panels to a group of reporters, he said that in a generation, they would "either be a curiosity, a museum piece, an example of a road not taken, or it can be a small part of one of the greatest and most exciting adventures ever undertaken by the American people; harnessing the power of the Sun to enrich our lives as we move away from our crippling dependence on foreign oil."[32]

Ronald Reagan beat Carter in the 1980 presidential race. And while Carter obsessed about energy, Reagan largely ignored the industry, figuring that it would manage itself. Reagan also worked to undo some of the regulatory obstacles that had been imposed on the industry by Congress and prior administrations. But the Great Communicator, like his predecessors (and successors), couldn't resist the allure of a familiar catch phrase. During his 1985 State of the Union address, Reagan declared that the U.S. should deregulate the natural gas market, "to bring on new supplies and bring us closer to energy independence."[33]

Reagan's successor, oilman George H. W. Bush, was prompted by Saddam Hussein's invasion of Kuwait to continue the foreign-oil-is-bad rhetoric.

On September 15, 1990, he spoke before a joint session of Congress, saying that the Iraqi invasion "helps us realize we are more economically vulnerable than we ever should be. Americans must never again enter any crisis, economic or military, with an excessive dependence on foreign oil and an excessive burden of federal debt." He went on to declare that Congress should immediately "enact measures to increase domestic energy production and energy conservation in order to reduce dependence on foreign oil."[34] In February 1991, just a few days before American and allied forces began their ground attacks against Iraqi military positions in Iraq and Kuwait during the First Iraq War, Bush laid out a new energy strategy. And his arguments were remarkably similar to those he had used in 1970 as a congressman arguing in favor of a quota on imported oil. On February 20, 1991, Bush said, "Our imports of foreign oil have been climbing steadily since 1985, and now stand at 42 percent of our total consumption. Too many of those oil imports come from sources in troubled parts of the world."[35]

Bush's solutions: Open the Arctic National Wildlife Refuge to drilling, ease regulations on new nuclear power plants, promote renewable energy, and ease regulations on natural gas pipelines and production. But Bush made the mistake of telling the truth about America's energy situation. "Let's talk about reality here," he said. In 1991, the U.S. was in a more secure position than it had been after the 1973 oil embargo, but, he continued, "We are, I will be the first to concede, a long way from total energy independence."[36]

During the 1992 presidential race, Bill Clinton's campaign machine seized on Bush's statement to declare that their man was an ardent believer in energy independence. And they insisted that Clinton could make America energy independent without drilling in the Arctic National Wildlife Refuge and without increasing the use of nuclear power. Clinton's campaign chairman in Texas, Garry Mauro, insisted that Clinton would lead America to the promised land of energy independence by increasing energy efficiency, pushing renewables, and increasing the use of "domestically abundant, clean-burning and inexpensive" natural gas. Mauro went on to declare that Bush and Reagan had "done

Atop the West Wing, June 20, 1979: Jimmy Carter shows the White House's new solar panels to reporters. Seven years later, during Ronald Reagan's administration, the panels were removed from the White House roof. In March 2007, one of the panels was placed on exhibit at the Jimmy Carter Presidential Library and Museum in Atlanta.
PHOTO CREDIT: Jimmy Carter Presidential Library and Museum.

nothing to halt the slide into the dependence abyss." He concluded by declaring that Clinton had "the foresight to make this country energy independent and the productive envy of the world."[37]

But during his eight years in office—aside from a failed effort to pass a carbon tax in 1993—Clinton pretty much ignored the energy business. And the "dependence abyss" grew just a little deeper. By the time the second George Bush declared the Second Iraq War in 2003, America's oil imports had grown from 42 percent of its total oil—the level that had prevailed during the First Iraq War and the first George Bush—to about 60 percent. (In 1973, when Nixon launched Project Independence, the U.S. was importing 35 percent of its oil.[38])

Jimmy Carter's solar panels ended up just as he predicted they might: as a museum piece. In 1986, Ronald Reagan had the panels removed from the White House while the roof was undergoing repairs. In March 2007, one section of the panels went on display at the Jimmy Carter Presidential Library and Museum in Atlanta as part of an exhibit regarding Carter's energy policies.[39]

The fact that those solar panels are now museum pieces shows that despite the upheavals in the global energy market that have occurred since 1973, neither the Congress nor the presidents who have served in the White House have been able to effect meaningful change in America's overall energy policy. The companies that used to dominate the oil business—Exxon Mobil, BP, Chevron, and Shell, to name a few— are no longer the power players in the industry. They have largely been replaced by the national oil companies. And the rise of the national oil companies is occurring at the same time that oil production in the non-OPEC countries is at, or near, its peak.

That peak in non-OPEC production, along with the surge in global energy demand, is the focus of the next chapter.

7

SEPTEMBER 15, 2004

The Non-OPEC Peak
Meets Surging Demand

> Oil remains fundamentally a government business. While many regions of the world offer great oil opportunities, the Middle East, with two-thirds of the world's oil and the lowest cost, is still where the prize ultimately lies.
>
> **DICK CHENEY, CEO OF HALLIBURTON, 1999**[1]

Since 1973, the global landscape has changed. Yes, OPEC continues to influence the global oil market. But today, the global energy market goes far beyond crude oil. The global trade in refined products like gasoline and jet fuel is accelerating. So, too, is the global trade in natural gas. The accelerating development of plants that produce liquefied natural gas means that gas-rich countries like Qatar, Algeria, Australia, and Nigeria can now ship their gas to almost any market on the planet.

The growing integration of the global market in everything from crude to liquefied natural gas is occurring at the same time that the non-OPEC energy producers—countries like Mexico, the U.S., the U.K., and others—are reaching, or have already passed, their peak output in oil and gas. The U.S. hit its peak oil output in 1970.[2] The U.K. and Colombia peaked in 1999. Norway and Oman peaked in 2000 and

Mexico in 2004. China will likely hit its peak in 2008, and Russia will likely hit its peak in 2012.[3]

These factors—the increasing global trade of all energy commodities, along with the peak in non-OPEC oil output—are coming to the fore at the same time that global energy demand is surging.

These trends were spelled out during a presentation to the Scottish Parliament in Edinburgh on September 15, 2004, by John Constable, an official from Exxon Mobil. After discussing the company's projections of global population increases (it expected a 27 percent increase by 2030) and oil demand growth (it expected demand to jump by about 50 percent, to about 115 million barrels per day by 2030), Constable got to the most important part of his presentation: peak oil production in the non-OPEC world.

Constable said that through 2010, the OPEC oil producers would not be able to increase their share of the global oil market. But, he quickly added, "Post 2010, the call [demand] on OPEC increases rapidly, requiring OPEC to add more than 1 MBD [million barrels per day] of capacity per year." He went on, saying that "the resources are adequate, and we expect that investments will be made in a timely manner to meet demand."[4]

That statement is important for three reasons. First, Exxon Mobil was not predicting the end of oil or a peak in global oil production. Second, it was a confirmation by one of the world's biggest (2006 revenues: $371 billion) and most conservative corporations that the global energy business—and the cash that comes with it—would increasingly be controlled by the members of OPEC.[5] Third, the report made it clear that energy independence—particularly when it comes to oil—is simply not realistic. The OPEC world will have effective control over the bulk of the world's most important transportation fuel for the foreseeable future, and there's nothing that the U.S. or any other country can do that will alter that fact.[6]

Similar work on the peak in non-OPEC oil production has been done by the clever analysts at John S. Herold Inc. (Unlike many other oil research firms, Herold does not do any brokerage business and

Table 1 Herold's Production Peak Estimates for the Supermajors

Company	Production Peak
Total S.A.	2010
Shell	2006
BP	2010
Exxon Mobil	2011
ConocoPhillips	2008
Chevron	2011
ENI	2008

SOURCE: John S. Herold Inc., 2006.

therefore doesn't have the conflicts of interest that plague much of the research done by Wall Street firms.) In late 2004, Herold began estimating when each of the world's biggest energy companies would reach their peak production. The analysts at Herold (who, by the way, were the first to call bullshit on Enron back in early 2001, several months before the company failed) believe that each of the world's seven largest publicly traded oil companies will begin seeing production declines no later than 2011.

In early 2007, Herold's CEO, Arthur L. Smith, told me that the firm had begun doing peak production estimates for individual oil companies for several reasons. Oil companies "are fighting depletion of their mature legacy fields," he said. And the companies that lack big new prospective fields under active development "are running fast and running scared. The estimated timing of peak production—and the severity of the rate of expected production declines—drives the strategies of oil companies between exploration, development, or acquisition. Meanwhile, the costs of finding and producing new oil and natural gas are surging beyond anyone's expectations. And finally, the volumes of new reserves from discoveries—that is, the reserve replacement rates—are substandard and getting downright depressing." (Smith left Herold in mid-2007.)

One of the key reasons for the rising cost of finding oil: the rising costs of raw materials like steel, cement, pumps, pipe, and other materials. The oil companies don't want to talk about when they will reach the limits of their production. Shell, Chevron, and other companies have refused to discuss Herold's work. The reasons are easily understood: CEOs make their living by promising ever better returns in the future. That won't work if the company's main product is running out and can't be replaced at a reasonable price.

Of course, no one has a crystal ball. Exxon Mobil and Herold could be flat wrong. And to be fair, since 2004, when Exxon Mobil first posited that 2010 would be the year that non-OPEC oil would peak, it has pushed that peak date back by a few years, thanks to new production from Canadian tar sands.[7] For its part, Herold has moved back some of the projected peaks for the large oil companies as those companies have found new reserves.

Nevertheless, it's clear that the "easy" oil has been found. The era of cheap energy has ended. Oil's getting harder and more expensive to replace. Over the coming decades, the biggest energy companies on earth will have no choice but to get more and more of their new energy supplies from OPEC and the national oil companies like Russia's Gazprom.

This shift in production toward OPEC and the national oil companies is occurring at the same time that demand in Asia is soaring. China provides the best example of this soaring demand. In 2005, Zhu Yu, the president of China's Sinopec Economics and Development Research Institute, said that between January and September of 2004, China's motor fuel use soared by 20 percent. Yu predicted that China's oil consumption would double by 2020 to more than 10 million barrels of oil per day. Meanwhile, the Indian economy continues to surge, and that country's oil consumption is expected to increase by nearly 30 percent by 2011 or so.

One of the key reasons for surging oil demand: car ownership in the Asia Pacific region, which is growing four times as fast as vehicle ownership in North America. In June 2007, Tata Motors, an Indian automaker,

Table 2a Ten Biggest Owners of Oil and Gas (combined)

Rank	Company	Reserves (billion barrels oil equivalent)
1	National Iranian Oil Co.	307.2
2	Saudi Aramco	305.6
3	Gazprom (Russia)	219.6
4	Iraq National Oil Co.*	135.1
5	Qatar Petroleum	128.9
6	Petroleos de Venezuela, SA	104.1
7	Kuwait Petroleum Co.	99.4
8	Abu Dhabi National Oil Co.	71.6
9	Nigerian National Petroleum Co.	40.2
10	Sonatrach (Algeria)	37.9

Table 2b Ten Biggest Owners of Oil Reserves

Rank	Company	Reserves (billion barrels)
1	Saudi Aramco	262.7
2	National Iranian Oil Co.	132.5
3	Iraq National Oil Co.*	115
4	Kuwait Petroleum Co.	89.4
5	Petroleos de Venezuela, SA	77.1
6	Abu Dhabi National Oil Co.	52.6
7	Libya National Oil Co.	28.8
8	Nigerian National Petroleum Co.	21.2
9	Lukoil (Russia)	15.9
10	Pemex (Mexico)	14.8

*Iraq does not have a national oil company. This name was created by Energy Intelligence in order to place Iraq's reserves in perspective.

SOURCE: Energy Intelligence, "The Energy Intelligence Top 100: Ranking the World's Oil Companies, 2006."

FIGURE 4 Anticipated Growth of Automobile Fleets in North America
and the Asia Pacific, 2000–2030

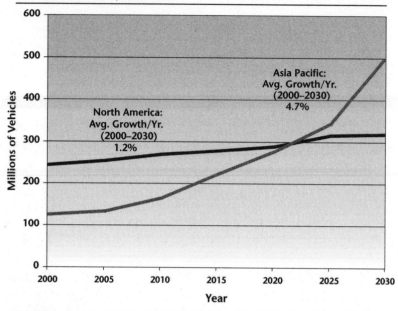

SOURCE: Exxon Mobil, 2005. These data come from two slides presented in
Exxon Mobil's 2005 Energy Outlook to 2030.

announced that it was planning to build a car for the Indian market that
would cost just $2,000. A few days after that news broke, Nissan, the Japanese automaker, announced it was thinking of building a $3,000 car for
the Indian market.[8] The flood of low-priced cars being built for the Indian market reflects the surging economic growth in the Asia Pacific region. By 2030, the Asia Pacific region will have nearly 500 million
vehicles. That's nearly a fivefold increase over the numbers seen in 2000.
And of course, all of those new vehicles will need fuel.

But the Asia Pacific region's thirst for energy goes beyond just gasoline and diesel. Demand for every type of energy commodity—natural
gas, uranium, coal, solar panels, and wind turbines—is also accelerating.

For instance, by 2030, the global consumption of natural gas is expected to more than double, much of that new demand coming from Asia Pacific countries. And much of the supply of natural gas will be coming from—who else?—OPEC and the national oil companies. Russia and the Middle East countries hold about 75 percent of the world's proven natural gas reserves.[9]

The point here is undeniable: Whether the commodity is oil or natural gas, the majority of future production will be coming from national oil companies—many of them members of OPEC. And there's nothing the U.S. can do to change that fact. Of course, some Americans don't like the idea that the U.S. cannot control its destiny when it comes to energy. They prefer the old days when the Texas Railroad Commission dictated the price of oil and the U.S. was largely autonomous in oil and natural gas. They don't like the increasing interdependence of the global energy business. And in particular, they don't like the idea that the Arab and Islamic countries of the Persian Gulf are going to continue to have a lot of influence over the future of the global energy market.

Those fears of the Arab world and the shifting balance of power have ignited a push by the neoconservatives to reframe the rhetoric used by Richard Nixon about energy independence. Instead of just protecting the American economy, the neoconservatives have turned energy independence into a catch phrase in the fight against terrorism. And in doing so, they have made the concept into one with broad appeal.

8

SEPTEMBER 27, 2004
The Neocon Crusade Begins

The most important event in the modern-day push for energy independence in America took place on September 27, 2004, at the National Press Club Building in Washington, D.C., when a group of America's leading neoconservatives launched their crusade against foreign oil. The featured speakers included three of the main neoconservative advocates for the Second Iraq War: Frank Gaffney, the head of a think tank called the Center for Security Policy; R. James Woolsey, the former director of the Central Intelligence Agency; and Ariel Cohen, an energy analyst from the conservative Heritage Foundation.

Those three men, along with other attendees and organizers of the meeting, have been the key provocateurs in the latest rhetorical push for energy independence. And looking back, it's clear that the 2004 meeting—which was sponsored by a new think tank called the Institute for the Analysis of Global Security, which was formed by another neocon, Gal Luft—was a milestone in the history of energy politics in America. By continually conflating oil and terrorism, Gaffney, Woolsey, and their fellow neocons have managed to push their message (and themselves) into the spotlight and gain credibility for the idea that energy independence is both needed and doable in order to keep America safe from terrorism.

Leading the rhetorical carpet bombing against foreign oil was Gaffney, who has long been among the most strident promoters of the claim that oil causes terrorism. America, he declared, is facing "a national emergency if not a full-fledged crisis. We are funding terrorism with our petrodollars. . . . We are paying them to kill us if we persist in this kind of behavior." And he added that "terror can have a crippling effect on our economy if this dependency [on Arab oil] persists."

Woolsey delivered a similar message. A native of Tulsa, Oklahoma, Woolsey has defined his post-CIA career by declaring that the U.S. is fighting "World War IV." And that war, Woolsey declared back in 2002, must be waged against the Shia in Iran, the Sunnis in Saudi Arabia, and the Baathists in Iraq and Syria. In the months before the Second Iraq War started, Woolsey—in language that would be echoed months later by George W. Bush—declared that those groups hate us "because of freedom of speech, because of freedom of religion, because of our economic freedom, because of our equal—or at least almost equal treatment of women—because of all the good things that we do." And he declared that "this is like the war against Nazism."[1]

During his speech at the National Press Club, Woolsey did not mention Hitler and the Nazis. But he did allow that the best way to cut America's consumption of Arab oil was by promoting alternative fuels. In particular, he declared that he wanted to see more ethanol production in the U.S. But Woolsey was apparently wary of the massive subsidies going into the corn ethanol business. And rather than be associated with that business, he added that while he wanted to see more ethanol production, he was emphatic that it be "not corn, not corn, not corn."[2]

Woolsey apparently understood that taking the side of Big Corn and its massive subsidies was not going to help his argument about the need for energy independence. He was insisting that ethanol should be made from other sources, like grass or wood chips. (It's worth noting here that by early 2007, Woolsey had changed his stripes with regard to corn ethanol. In his March 2007 speech to the Virginia farmers, Woolsey said that "American farmers, by making the commitment to grow more corn for ethanol, are at the tip of the spear on the war against terrorism."[3])

Gaffney and Woolsey were the stars of the program. And that was just fine with them. Preaching about the evils of foreign oil and the crying need for energy independence gave them a new cause. After the fall of Saddam Hussein, the neoconservatives needed a new bogeyman, and Arab oil provided the perfect foe on which they could focus their fire. Both Gaffney and Woolsey were staunch backers of the Second Iraq War, and both had been among the earliest advocates for the invasion of Iraq. The two men were among the first to join the Project for the New American Century, the group of hawks who began pushing for the overthrow of Saddam Hussein in 1998.

Gaffney was one of the signatories of the Project's original statement of principles, released in June 1997, which advocated increased military spending and a "Reaganite policy of military strength."[4] Gaffney's cosigners of that original document included:

- Dick Cheney, who would become vice president under Bush.
- Donald Rumsfeld, who would become Bush's Defense Secretary.
- I. Lewis "Scooter" Libby, who would become Cheney's chief of staff and a member of the White House Iraq Group, which led the media relations effort to sell the Iraq war to the American public in the months prior to the invasion.[5] (In June 2007, Libby was sentenced to 30 months in prison and fined $250,000 for lying and obstructing an investigation related to the leak of Valerie Plame Wilson's identity as a CIA agent.[6])
- Eliot Cohen, an academic from the Paul H. Nitze School of Advanced International Studies at Johns Hopkins University, who in 2007 became a top adviser to Secretary of State Condoleezza Rice.[7]
- Midge Decter, journalist and the wife of leading neocon Norman Podhoretz.[8] In 2003, Decter authored *Rumsfeld: A Personal Portrait*. A review by *Publishers Weekly* called it "an adoring biopic from an adoring writer" that was "thoroughly uncritical."[9]

Woolsey was a signatory to the now-famous January 26, 1998, letter sent by the Project for the New American Century to President Bill

Clinton, which declared that U.S. strategy should aim "above all, at the removal of Saddam Hussein's regime from power."[10] Woolsey's fellow signers on that letter included other high-profile neocons who would go on to become officials in George W. Bush's administration, including:

- Richard Perle, the man known as the "Prince of Darkness" who would become chairman of the Defense Policy Board under Bush and who continues to be one of the most prominent neoconservatives.[11]
- John Bolton, who would become U.S. ambassador to the United Nations.[12]
- Rumsfeld.
- Robert Zoellick, an adviser to Enron who would become U.S. trade representative and Deputy Secretary of State under Bush.[13] (Zoellick is now the head of the World Bank.)

After signing the 1998 letter, Woolsey continued his cheerleading for an American invasion of Iraq. And his rhetoric went into overdrive after the September 11, 2001, attacks. In December 2001, he told the *Washington Post* that there was no doubt that Saddam Hussein was developing biological and nuclear weapons. *Post* reporter Nina J. Easton wrote that "Woolsey lays the groundwork of his case with what he calls 'the slam dunks.'" She then quoted Woolsey directly: "There is so much evidence with respect to his development of weapons of mass destruction and ballistic missiles that I consider this point beyond dispute." He continued, asking rhetorically, "Has the Iraqi government been involved with terror, especially terror against the United States? Again, this one is a slam dunk." Woolsey told Easton that "we need to destroy this regime that wants to destroy us."[14]

Among the other speakers at the September 2004 program at the National Press Club was yet another neocon: Ariel Cohen, a pro-Iraq-war booster from the conservative Heritage Foundation. In the months before the war, Cohen repeatedly argued that the U.S. needed to take action against Saddam Hussein. If it did not, warned Cohen, then Saddam could use oil as an economic weapon. In a September 2002 speech, Cohen claimed that if left in power, Saddam "may drive oil prices *as high as $45–$50, plunging the world into unprecedented recession*" (italics added).[15] In December 2002, Cohen, along with another Heritage Foundation operative, Gerald O'Driscoll, wrote a piece for the *National Review Online* in which they advocated an invasion of Iraq and the privatization of its oil sector as a way to weaken OPEC's hold on the global oil market. They declared that "Iraq's privatizations of its oil sector, refining capacity, and pipeline infrastructure could serve as a model for privatizations by other OPEC members, thereby weakening the cartel's domination of the energy markets."[16]

The program at the Press Club was coordinated by Luft, who received his doctorate in strategic studies from the Nitze School at Johns Hopkins. The Nitze School has long been a hotbed of neoconservatism. Its roster of academics includes Eliot Cohen and Francis Fukuyama, who, like Cohen, was an early member of the Project for a New American Century.[17] (Fukuyama has since left the fold. In 2006, he wrote a book in which he declared that neoconservatism "as both a political symbol and a body of thought, has evolved into something that I can no longer support."[18]) The former dean of the Nitze School is Paul Wolfowitz, another founding member of the Project for a New American Century. He went on to become Deputy Defense Secretary from 2001 to 2005.[19]

At the time of the meeting, Luft's new think tank was getting positive publicity thanks to its efforts to track pipeline attacks in Iraq. It was also getting attention because Luft had been conflating oil and terrorism for more than a year. In March 2003, he gave a speech in York, Pennsylvania, in which he declared that "America's dependence on foreign oil is helping to feed billions to the beast that is trying to destroy us."[20] In fact, Luft was regularly conflating oil and terrorism while using every opportunity to declare that the U.S. had to quit buying foreign oil and increase the use of renewable and alternative fuels. Just a few days before the conference at the National Press Club, Luft told the *Sunday Times* of London that "every new car should have a mandatory flexible fuel capacity," meaning it should be able to burn ethanol as well as gasoline.[21]

About three dozen people attended Luft's conference. The main news generated by the event was the unveiling of the Set America Free Manifesto. The manifesto declares that the U.S. can "no longer . . . postpone urgent action on national energy independence." It goes on, saying that "over the next four years" the U.S. should achieve "a dramatic reduction in the quantities of oil imported from unstable and hostile regions of the world."[22] (For the full text of the manifesto, see Appendix B.)

The signers of the declaration included Gaffney and Woolsey, as well as a who's who of America's prowar neoconservative leadership. One of the signers was the Hudson Institute's Meyrav Wurmser. In 1996, she was one of the authors—along with Perle and Douglas Feith—of a strategy paper for Israeli prime minister Benjamin Netanyahu that called for the overthrow of Saddam Hussein and for military assaults against Lebanon and Syria. Wurmser's husband, David Wurmser, was also an author of that report.[23] He went on to become a top adviser to Vice President Dick Cheney.[24]

Woolsey, Gaffney, Luft, and their fellow travelers who signed the Set America Free petition succeeded in creating a self-perpetuating neoconservative echo chamber. That echo chamber allowed the various signatories to sit on each others' boards and tout each others' positions on energy independence and thereby make their "movement" appear larger than it actually was. For instance, the advisory members at the Institute for the Analysis of Global Security include many of the original signers of the petition, including Woolsey and Robert C. McFarlane, the former National Security Adviser under Ronald Reagan who, in 1988, pled guilty to four counts of withholding information from Congress for his role in the cover-up of the Iran-Contra scandal.[25] Luft's advisory group also includes the Nitze School's Eliot Cohen.[26]

The neocons' echo chamber included two other neoconservative groups: Gaffney's Center for Security Policy and the Committee on the Present Danger, a Cold War–era outfit that was relaunched in July 2004, just a few weeks before the Set America Free meeting at the National Press Club. The committee's new purpose: to "advocate strong policies both against international terrorists and their sponsors and in favor of freedom and security."[27]

Since the meeting at the Press Club, the neocons' crusade for energy independence has garnered an enormous amount of attention. By early 2007, the Set America Free Manifesto had been signed by a number of other groups, including the Natural Resources Defense Council, which is one of America's biggest environmental groups; the Apollo Alliance, an alliance of environmental and labor groups; Felix Kramer of CalCars, a group that promotes hybrid vehicles; and Tom Daschle, the former minority leader in the U.S. Senate, who, by early 2007, was the cochair (along with another longtime ethanol booster and former senator, Bob Dole of Kansas) of a proethanol outfit called the 21st Century Agriculture Policy Project.[28]

Given the diversity of the groups supporting the Set America Free Manifesto, Woolsey began calling it a coalition of "tree huggers, do-gooders, sodbusters, cheap hawks, and evangelicals."[29] And the echo chamber technique was working brilliantly. The twin mantras of energy independence and oil-causes-terrorism were being repeated over and over. In 2005, national newspapers carried nearly four dozen stories that contained "energy independence," "oil," and "Set America Free." Between late 2004 and early 2007, Gaffney and Woolsey had been mentioned or quoted in more than 200 articles in newspapers and magazines discussing the issues of oil and terrorism.[30] Gaffney regularly used his column in the *Washington Times* to harangue his readers about the evils of the petrostates in the Middle East. And Woolsey used every opportunity to talk about energy independence. For instance, on January 1, 2007, Woolsey wrote a piece for the *Wall Street Journal Online*. Typically, he attacked both the Saudis and the Iranians. And true to form, he also declared that Americans should "bet on major progress toward [energy] independence."[31]

But Woolsey and the rest of the Set America Free crowd are assiduously ignoring facts that don't fit their arguments. Part 3 will show that the neocons' fantasy about energy independence is remarkably similar to the intelligence that they used to justify the Second Iraq War: a lot of innuendo, a whole lot of bluster, but precious few facts.

PART THREE

THE IMPOSSIBILITY
OF INDEPENDENCE

9

SKEPTICISM AND THERMODYNAMICS

When it comes to discussions about energy and energy policy, the overwhelming majority of Americans lack two key attributes: sufficient skepticism and sufficient knowledge of basic science.

Let's discuss skepticism first. Consider the case for wind power. According to a 2007 survey, 90 percent of Americans support building more wind power capacity.[1] Many of those respondents probably believe that given enough wind turbines, America's electricity needs will be solved and the U.S. will see a drastic reduction in carbon dioxide emissions.

There are plenty of boosters eager to hype wind power. For instance, Greenpeace claims that "wind could supply more than three times the total amount of electricity currently produced in the United States."[2] In Colorado, the upscale ski area Vail Resorts Inc. claims that its chairlifts and lodges are "100% powered by wind." The resort reportedly buys renewable energy certificates from a Colorado company that promises it is somehow, somewhere, buying electricity that it claims is renewable.[3] George W. Bush has chimed in on wind. In 2006, he claimed that wind power could provide 20 percent of America's electricity needs.[4] Another wind booster is *New York Times* columnist Thomas Friedman, who in late 2006 wrote a column in which he said the U.S. should build a "national electricity transmission grid from the Dakotas to Texas to take wind electricity from where it is best produced to the big cities where it is most needed." Once that's done, Friedman says, wind power can be used to power hybrid electric cars. And if that works, he claims, "It could be better than Kyoto," implying that the results from such a scheme would mean dramatic reductions in the overall output of carbon dioxide that could surpass those called for in the Kyoto Protocol.[5]

Skepticism is in order. While there's no question that wind power production is growing rapidly, many people confuse installed capacity for actual power production. But even in the best locations, wind turbines produce power only about one-third of the time. And many produce at lower rates. Wind power advocates always prefer to talk about the total generating capacity of their wind farms, which are usually measured in megawatts. But the key measure for electricity is kilowatt-hours. And that's where a skeptical look at wind power gets rewarded.

By 2010, the U.S. will be generating only about 50 billion kilowatt-hours per year from wind power.[6] That's a pittance compared to the 800 billion kilowatt-hours of electricity that are pumped out every year by America's 104 nuclear power reactors.[7] That quantity of wind power looks yet smaller when compared to America's 400-plus coal-fired power plants, which generate more than 2,000 billion kilowatt-hours of electricity per year.[8]

I provide this example not to pick on wind power, but to provide a sense of scale and to make the point that renewable and alternative energy sources face a long, hard slog in the effort to displace significant quantities of fossil fuels. And that's not because of any scandalous plot launched against them by Big Oil or evil utility executives. Rather, it's that wind energy, solar energy, and biofuels are all fighting the laws of thermodynamics.

Which brings us to the next point: Precious few Americans understand the rules of the energy game, and that lack of scientific knowledge allows them to be easily gulled by politicians, environmental groups, and others who want to advance a no-pain energy agenda that sounds appealing but has no basis in reality.

The energy business is ruthlessly policed by the immutable laws of thermodynamics. The first law of thermodynamics says that energy is neither created nor destroyed. The chemical energy that is stored in the jet fuel inside a Southwest Airlines jetliner in Dallas is, during a flight to El Paso, turned into heat in the atmosphere or heat on the tarmac, or perhaps into air-conditioning that keeps the passenger in seat 8A comfortable. Thus, the energy in the jet's fuel tank doesn't disappear; it gets redistributed. And that redistribution of concentrated energy to a more random form leads us to the second law of thermodynamics, which says that energy tends to become more random and less available. Heat always dissipates, from hotter to colder, never the other way around. An air-conditioned room during the summer in Kuwait City or a heated room in Copenhagen during the winter will quickly turn uncomfortable unless more conditioned air is pumped into it. And creating that conditioned air, whether it's cooled or heated, takes a steady flow of energy.

Taken together, the first two laws of thermodynamics provide the key to understanding why fossil fuels are so dominant in today's economy. Turning diffused sources of energy—whether that energy is stored in the starch found inside corn kernels, the photons in sunlight, or the kinetic energy available from a strong breeze blowing off Cape Cod—into more concentrated forms of energy is always an uphill climb. And the more diffused the energy source, the more difficult it is to concentrate it into a form that can provide usable work, whether that's lighting a house, powering a car, or firing a furnace.

For example, it takes a lot of energy to convert a field full of Iowa corn into viable motor fuel because that corn has to be planted, irrigated, fertilized, harvested, and refined. And each step in that planting, nurturing, harvesting, and refining process requires energy. Given that heat energy tends toward randomness and disorder, the more steps that are involved in that energy concentration process, the less efficient the process becomes. That point can be grasped more easily by looking at the heat value of fossil fuels. Or to be more specific, by looking at their energy densities.

Crude oil contains about 18,400 Btus per pound and coal contains about 10,400 Btus per pound.[9] Now compare those heat values with the main feedstocks for ethanol. Corn contains 7,000 Btus per pound.[10] And switchgrass—a fast-growing plant that's native to North America and is often mentioned as a prime candidate for producing cellulosic ethanol—contains just 6,400 Btus per pound.[11]

Obviously, the higher energy density of crude oil (nearly three times as much as that found in switchgrass) gives it a major thermodynamic advantage. In fact, the lightest grades of crude can almost be pumped straight from the oil well into an automobile tank. Further, crude oil is already in liquid form, so it is far easier to transport and handle during the refining process. By contrast, both corn and switchgrass must be mixed with large quantities of water, fermented, and then distilled into a usable liquid form before they can be utilized. And each of those steps takes energy.

The purpose of this tutorial on thermodynamics is to reinforce one of the themes of this book: the essential role that fossil fuels will play in the world's energy mix over the coming decades. While politicians and environmental advocates may wish for a different paradigm, the laws of thermodynamics, combined with the many trillions of dollars that have been invested in fossil-fuel-powered automobiles, airplanes, power plants, industrial systems, and residential heating and cooling systems, will mean that fossil fuels will predominate for the foreseeable future.

Those facts provide the background for Part 3. These same facts are also essential reasons why America cannot be energy independent and why the attempt to transition from fossil fuels to alternatives and renewables will be difficult and lengthy.

Part 3 is designed to give readers the unvarnished truth about America's ever-growing appetite for energy, the limited effect that efficiency will have on overall energy consumption, and the myriad problems associated with ethanol production. It looks at the many challenges faced by other energy sources, including wind and solar, while also showing why America will need to continue importing fuels like oil, natural gas, coal, and uranium.

10

THE BEST BUY EFFECT

Americans love their gadgets. And electronics retailer Best Buy Co. Inc. operates over 1,000 stores to keep consumers supplied with those gadgets—nearly every one of which comes with a power cord. The plethora of power cords that has been sold by Best Buy over the past few years has led to a corresponding surge in demand for electricity. Between about 1998 and 2006, the amount of electricity that American homes used to power their consumer electronics—TVs, home theater systems, answering machines, VCRs, cordless phones, computers, and printers—*more than doubled.*[1] In 2006 alone, consumer electronics in the U.S. consumed about 147 billion kilowatt-hours of electricity.[2] That's more electricity than was produced by all of the windmills, solar panels, and biomass-to-electricity plants in America that year.[3]

This surging demand for electricity might be called the Best Buy Effect. Americans' love of technology and high-tech toys has led to a corresponding increase in electricity demand that shows no sign of slowing, regardless of any discussions about peak oil, energy independence, or global warming.

Of course, consumer electronics represent only a portion of the overall electricity use in America. And the ever-growing amounts of electricity needed to power these devices leads us to a rather obvious point: Americans love energy. They don't love energy itself; they love it for what it does. And over the past few years, Americans have fallen in love with bigger and bigger flat-screen TVs, ever-more-powerful personal computers, and fancier, more complicated home entertainment systems—every one of which requires more and more electricity. For instance, large flat-screen plasma TVs consume up to four times as much electricity as the older, smaller, cathode-ray-tube models.[4]

All of these new electronics have helped continue to drive electricity demand ever upward. In mid-2007 electricity demand was growing by 2.7 percent per year. If demand continues growing at that rate, electricity consumption in the U.S. will double in about 26 years.[5] By late 2006, electricity demand was growing so rapidly that the North American Electric Reliability Council warned that by 2015, the U.S. may have a shortfall of 81,000 megawatts of generating capacity.[6] Just to put that number in perspective, the Three Mile Island Generating Station in Pennsylvania—the most famous nuclear power plant in America—operates one nuclear reactor, which produces 850 megawatts of electricity.[7] Thus, the looming electricity shortfall in the U.S. is equal to the output of 95 nuclear reactors.

It's not just electricity. Americans also love gasoline. And the upward trends in motor fuel use mirror those seen in electricity consumption.

Americans are incredibly mobile. They also love to fly, ride motorcycles, and take boating trips. But they really love their cars. In 2004, according to the Bureau of Transportation Statistics, American motorists drove more than 2.9 trillion miles—that's nearly double the distance that Americans drove in 1980 and nearly triple the amount driven in 1970.[8] Further, the number of miles that Americans drive has increased every year since 1960.[9] Given these statistics, it's hard to imagine anything (other than drastically higher fuel prices) that will convince Americans to drive less.

Americans not only love to drive, but they also love to drive fast. And driving fast requires lots of horsepower. As one auto writer put it, "There's no such thing as too much sex, or too much horsepower."[10]

Indeed, bigger vehicles and more horsepower are key reasons behind America's ever-growing motor fuel use. In 1975, the average American car (or light truck) weighed 4,060 pounds. By 1987, it weighed just 3,220 pounds. But by 2004, it had surged to 4,066 pounds. In 1975, the average American vehicle had 137 horsepower. By 1987, that power output had dropped to 118 horsepower. But by 2004, horsepower had jumped to an average of 208. As one reporter for *Environment* magazine explained, "These trends have had a negative impact on fuel economy far more dramatic than the impact of increasing SUV sales."[11]

This ever-rising desire for horsepower was on display during the 2007 Detroit Auto Show, at which Chrysler unveiled a new Viper sports car with 600 horsepower.[12] One auto writer at the show declared that cars with horsepower "ranging up to 400 and beyond are no longer so rare."[13] That could be seen in the pickup truck business, where the various automakers have engaged in a horsepower arms race. In 2007, Toyota was offering a Tundra pickup with a 381-horsepower V-8 engine.[14] Chevrolet, meanwhile, was offering Silverado pickups stocked with a 367-horsepower V-8.[15] The horsepower obsession was obvious in the passenger car segment as well. The German automaker BMW was selling the 760Li luxury sedan, equipped with a monster V-12 engine producing 438 horsepower.[16] Two other German automakers, Porsche and Audi, were both offering American drivers sports cars with more than 415 horsepower.[17]

Over the past two decades, car retailers say that among consumer desires, fuel efficiency has consistently ranked behind other amenities like cup holders and sound systems. That point was made clear in April 2007, when the CEO of AutoNation, America's largest publicly traded auto dealer, told the *Wall Street Journal* that "you have to look past what the consumer says they are going to do and the moment where they write a check. . . . That's the moment of truth. They want size and speed."[18]

This push for more horsepower has had a predictable result: America's auto fleet is scarcely more efficient today than it was immediately after the 1973 oil embargo. That lack of efficiency has helped drive up fuel consumption. Nearly every year, gasoline demand in the U.S. goes up. For instance, between 1983 and 2005, gasoline consumption rose by almost 32 percent.[19] And the Energy Information Administration expects America's gasoline consumption to continue rising. In 2005, the U.S. consumed 140 billion gallons of gasoline. By 2030, the EIA expects gasoline consumption to hit 192.1 billion gallons.[20]

The agency's projections of America's future gasoline consumption and electricity consumption (expected to increase by about 43 percent by 2030) fit perfectly with the pattern seen over the past six decades or so.[21] And that pattern shows a near-constant increase in total energy consumption in the American economy. And as U.S. energy consumption has increased, so, too, have productivity, employment, and economic output. All of which can be seen in Figure 5.

As the graphic makes clear, there is a strong correlation between energy consumption, productivity, employment, and economic growth. This rising energy consumption is part of an inexorable trend: The

FIGURE 5 Productivity, Energy, and the Economy in the U.S.

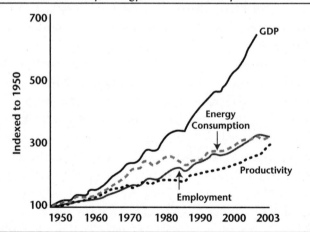

SOURCE: EIA, Annual Energy Review 2003; Department of Commerce, Department of Labor, Bureau of Labor Statistics.

Image is courtesy of Huber and Mills, *The Bottomless Well*, and www.digital powergroup.com, http://digitalpowergroup.com/TBW/downloads/figure81.pdf.

more energy we use, the more energy we want. Energy consumption is akin to bandwidth on the Internet, or sex, or horsepower, or clock speed on computer processors: The more we get the more we want.

This trend can also be attributed to the wealth effect. As people get richer, they want more stuff—whether it's a fishing boat, a bigger house, a bigger television, or a new freezer doesn't matter. Each of those things results in greater energy use. And the wealth effect is true all around the world. As people get richer, they use more energy. As soon as a family get a nice house, they want to heat it in the winter; then comes air-conditioning for summer. Fuel use begets more fuel use. And this is true always, everywhere.

The advent of the Internet and the Internet economy provides another example of how energy use begets more energy use. It's a phenomenon that I call the Google Effect.

About a decade ago, there was furious debate over how much electricity was used by the Internet. Some efficiency experts claimed that the rise of the World Wide Web would mean less energy use because more people would be able to telecommute. But a pair of sharp energy analysts, Peter Huber and Mark Mills, made the opposite claim. They said that the Internet would lead to more electricity use, not less. And it appears they were right.

Everyone knows about Google. Founded in 1998, the Internet search company has become one of the most valuable and most profitable companies on earth.[22] But the Google phenomenon is based entirely on cheap electricity and lots of it. In 2001, the company had about 8,000 computer servers to manage the millions of Web pages that it tracked. Two years later, Google had 100,000 servers. By mid-2006, the company had more than 450,000 servers spread among some two dozen locations around the world. Each one of those computing locations—known in the industry as data centers—consumes huge amounts of electricity, both to power the computers and to run the air conditioners needed to keep the computers from overheating. And Google is building ever larger data centers. In 2006, it built a massive data center in The Dalles, Oregon, that covers an area the size of two football fields. The

data center's heat problems were so extreme that Google even built two
four-story-high cooling towers to dissipate the heat generated by the
servers.[23] Although Google doesn't disclose how much power the data
center uses, it likely uses as much power as 11,000 homes.[24]

Google provides just one example of how the booming Internet
economy has resulted in higher energy consumption. Other companies
like Yahoo! and Microsoft—which may have 800,000 computer servers
operating by 2011—are also using huge quantities of electricity in order to keep their servers running.[25] In February 2007, a scientist at
Lawrence Berkeley National Laboratory released a study showing that
in 2005, computer servers consumed about 1.2 percent of all the electricity in the U.S., or about as much as all of the color TVs in America.
More important, perhaps, is this fact: Between 2000 and 2005, the
amount of electricity used by those servers doubled and now equals
about 5 gigawatts.[26] (A gigawatt is 1,000 megawatts.)

The computer servers in America now consume nearly six times as
much electricity as can be generated by the reactor at the Three Mile
Island nuclear power plant. And it is likely that the growth in server
capacity will continue as larger volumes of data are exchanged via the
Internet. The phenomenal popularity of online video sites like YouTube,
the advent of podcasting, and the increasing delivery of video via the
Web mean that Internet users are accessing larger data files than ever
before, and all of those data must be stored on ever-increasing numbers
of servers.

The massive data centers operated by companies like Google, Yahoo!, and Microsoft are the Information Age equivalents of aluminum
smelters or steel plants—massive industrial facilities with huge ap-
petites for electricity. But instead of turning out metal, they churn out
torrents of data.

The growing hunger for electricity to drive the Internet can also be
seen by looking at semiconductors. In the early 1990s, one of Intel's
first Pentium processors used about 10 watts of power.[27] By 2006, Intel's Xeon processor was using up to 165 watts of power.[28] As computers have become more powerful in processing information, they have
also gotten ever hungrier for electric power.

FIGURE 6 Increases in Microprocessor Peak Power Consumption (in Watts)

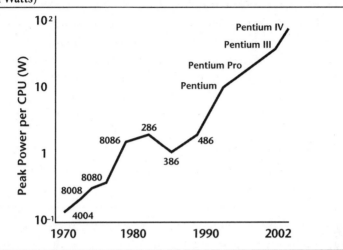

SOURCE: Intel.

Image is courtesy of Huber and Mills, *The Bottomless Well*, and www.digital powergroup.com. Images available: http://digitalpowergroup.com/TBW/downloads/figure25.pdf.

This combination of the Best Buy and Google Effects, along with Americans' desire for more horsepower in everything from their cars to their computers, is driving energy use ever upward. That trend can be seen from the last few decades. Over the 25-year period from 1980 to 2004, global electricity consumption more than doubled.[29] And over the coming decades, electricity and oil consumption will almost certainly continue climbing, regardless of any gains that are made in energy efficiency. That upward trend will likely be driven by increasing hunger for energy of all types in developing countries.

A case in point: In March 2007, at South by Southwest Interactive, part of Austin's massive South by Southwest festival, a panel discussion focused on an intriguing topic: "looking for the next billion Internet users." A focal point of the discussion was Africa. While the panelists mostly focused on the benefits that Internet connectivity will bring to Africans living in remote areas, scant time was spent dis-cussing the electricity needed to power the computers needed to create those benefits. While several companies, including Advanced Micro Devices, Intel, Google, and others, are looking for ways to make com-puters use less electricity, even the most efficient machines require about 17 watts of power. If it can be assumed that every one of those 1 billion new Internet users will require just 17 watts, and that there will be no servers tying those users to information (which, of course, will not be the case), those 1 billion new Web surfers will need at least 17,000 megawatts of new power plants. That's as much electricity as the amount that would be produced by 20 nuclear reactors like the one at Three Mile Island.

The point is clear: America's experience of rising income and rising energy use will be duplicated by other countries around the world. Several billion people around the world are eager to start Googling and watching big-screen, high-definition TVs. They also want cars, air-conditioning, computers, and refrigerators. As they begin buying those machines, electricity demand, and oil demand, will rise.

The result of this rising demand for energy in the U.S. and else-where is equally obvious: Unless the U.S. mandates that consumers drastically cut their energy consumption, America's dependence on foreign sources of energy of all types—oil, coal, uranium, natural gas, and electricity—will increase, not decrease. And that rising demand will mean that energy independence is little more than a pipe dream.

11

MORE EFFICIENCY, MORE FUEL

Nearly all of the discussions about energy independence and global warming include prominent mentions of the need for better energy efficiency.

The argument is simple: If only we use energy more efficiently, consumption will fall and everything will be better. There's no doubt that efficiency is a marvelous thing. It allows consumers to get more work out of the same pound of coal, or gallon of gasoline, or windmill blade, or photovoltaic cell. And the more efficient a given process becomes, the more profitably it can be used. A car that gets 30 miles per gallon can effectively deliver much of the same value as one that gets 15 miles per gallon—and do so at half the fuel cost. A compact fluorescent light bulb that consumes 18 watts of electricity and yet delivers the same amount of light as an incandescent bulb using 60 watts makes a great deal of economic sense.

But efficiency alone won't deliver energy salvation. Proof comes from other technological innovations. In the early days of the personal computer, there were claims that the computer would result in the advent of the paperless office. That didn't happen. Instead, whole new industries, like desktop publishing, were born, resulting in ever greater amounts of paper consumption.[1] Likewise, predictions that greater efficiency would result in lower energy consumption have proven utterly and completely wrong.

For decades, "green" energy promoter Amory Lovins has been claiming that greater efficiency would lower energy demand. For instance, in 1984, Lovins told *Business Week* magazine that "we see electricity demand ratcheting downward over the medium and long term. The long-term prospects for selling more electricity are dismal. . . . We will never get, we suspect, to a high enough price to justify building centralized thermal power plants again. That era is over."[2]

Except that it isn't.

America's electricity production jumped by about 66 percent since Lovins made his declaration, rising from 2,400 billion kilowatt-hours in 1984 to just over 4,000 billion kilowatt-hours in 2005.[3] And to meet that demand, utilities have built dozens of centralized thermal power plants. In fact, Lovins has repeatedly been proven wrong when it comes to energy trends. In 1976, he predicted that renewable energy would be supplying about 30 percent of the total energy demand in America by 2000. The reality, when excluding hydropower, was closer to 1 percent. And yet, "Inexplicably," notes Vaclav Smil of the University of Manitoba, "Lovins retains his guru aura no matter how wrong he is."[4]

Just as Lovins wrongly predicted that efficiency would quell electricity demand, there is a widespread belief that federal mandates for higher-mileage cars will result in less fuel consumption. In September 2005, after Hurricane Katrina caused fuel supply problems in the southern U.S., the *New York Times* published an editorial concluding that the U.S. cannot drill its "way out of oil dependency and high prices. The only sure relief will come through improved fuel efficiency."[5]

The *Times*'s editorial board may be convinced, but there's precious little evidence to prove that fact. History shows that as the U.S. economy has grown more energy-efficient, energy consumption has continued climbing. In 1980, the U.S. was using about 15,000 Btus per dollar of GDP. By 2004, the energy intensity of the U.S. economy had improved dramatically so that just over 9,000 Btus were required for each dollar of GDP.[6] The EIA expects those efficiency gains to con-

FIGURE 7 Energy Intensity of the U.S. Economy, 1950–2010

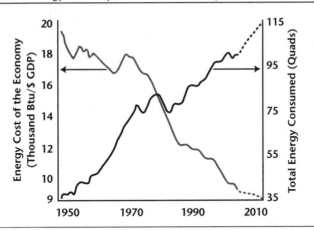

The line on the top left shows that the amount of energy needed to produce goods and services has been on a steady downward trend. In 1950, it took about 19,000 Btus to produce $1 of gross domestic product. By 2010 or so, the U.S. will need only about 9,000 Btus to produce a similar amount of output. Also, note that a "quad" is a common energy measurement. One quadrillion Btus is equal to about 172 million barrels of crude oil. http://www.eia.doe.gov/neic/infosheets/apples.html.

Sources: EIA, Bureau of Economic Analysis.

Image is courtesy of Huber and Mills, *The Bottomless Well*, and www.digital powergroup.com. Available: http://digitalpowergroup.com/TBW/downloads/figure72.pdf.

tinue. The agency projects that by 2030, energy intensity will fall from about 9,000 Btus per dollar of GDP to about 5,800 Btus per dollar of GDP. But even with that dramatic increase in efficiency, overall energy consumption in the U.S. will rise by more than 30 percent, from 100.1 quadrillion Btus in 2005 to 131.1 quadrillion Btus in 2030.[7]

What's true of the broad economy is also true of automobiles. Toyota Priuses and other hybrid cars are cool. But they are, as one Houston-based oil industry analyst put it, a "Band-Aid on an amputee." Even dramatic increases in America's automobile fuel efficiency will likely only

FIGURE 8 Projected Growth in Daily U.S. Oil Demand by 2025 Under Various Fuel Economy Scenarios

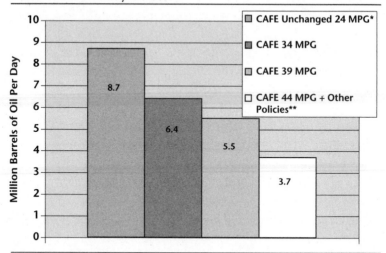

SOURCE: National Commission on Energy Policy, 2004.

*These data combine car and light truck fuel economy standards.

**These data include higher standards for heavy-duty tractor trailers and policies to promote renewable fuels.

slow the rate of growth in the amount of oil we are importing from abroad. In late 2004, a group of Washington power brokers and insiders, calling themselves the National Commission on Energy Policy, looked at the Corporate Average Fuel Economy (CAFE) standard, the federal mandate that requires the automakers to meet certain efficiency targets. The group determined that even if Congress mandated that the domestic auto fleet increase its average fuel economy to 44 miles per gallon—a major increase over the 27.5 miles-per-gallon standard in effect in 2007—America's motor fuel consumption will still increase by 3.7 million barrels per day by 2025.[8]

Indeed, America's motor fuel consumption continues ever upward. For instance, in February 1983, the U.S. was using about 259 million gallons of gasoline per day. By February 2007, that figure had jumped by 44 percent to nearly 373 million gallons per day.[9]

There are a number of reasons why American motorists are using more fuel.

First and foremost, Americans have a lot of machines that burn motor fuel. In 2005 (the last year for which statistics are available), the U.S. had 247.4 million registered motor vehicles. That's more than double the number of vehicles that were on American roads in 1970. In addition to the huge number of vehicles, Americans owned over 224,000 general aviation aircraft and 12.9 million recreational boats.[10] And, of course, those numbers don't count the proliferation of other machines that use motor fuel—snow blowers, generators, tractors, lawnmowers, and chainsaws, to name but a few.

Second, Americans are keeping their vehicles longer, so that older, less efficient cars will be staying on the road for substantially longer periods. In 2005, the average motor vehicle in the United States was 9 years old. That's a big jump when compared to 1990, when the average vehicle was just 6.5 years old.[11] People are keeping their cars longer for a simple reason: Today's cars are much higher quality than they were two decades ago.

Third, given America's huge motor fleet and its age, replacing it with a more fuel-efficient fleet will take decades. This long lag time between the scrapping of older cars in favor of newer, more efficient ones is often overlooked. It simply doesn't make sense for most consumers to get rid of their current vehicle—even if their fuel bills are relatively high—to replace it with a more efficient one.

While consumers pay homage to the Toyota Prius and other super-efficient hybrid cars, they are still buying SUVs and pickups that use lots of fuel. In 2005, the number of hybrid vehicles sold in America doubled to about 200,000. That same year, hybrids were outsold by SUVs by a ratio of 23 to 1.[12] In 2006, hybrid sales continued their upward trend, increasing by 28 percent over the year-earlier numbers. But even with that increase, hybrids still accounted for only about 1.5 percent of all the cars sold in America.[13] Those sale numbers show that

American drivers love the concept of energy independence and hate the fact that the U.S. buys foreign oil. But when it comes time to strap on their seat belts, they aren't as interested in efficiency as they are in the comfort, convenience, and power offered by larger vehicles.

The limits of energy efficiency were made clear by Peter Huber and Mark Mills in their provocative 2005 book on the energy business, *The Bottomless Well*. The two concluded that "efficiency doesn't lower demand, it raises it." They explained that the pursuit of energy efficiency has been the "one completely consistent and bipartisan cornerstone of national energy policy since the 1970s." And yet, even though overall energy efficiency has increased dramatically since that time, "demand has risen apace."[14] This passage explains why energy demand will almost surely continue rising:

> Efficiency may curtail demand in the short term, for the specific task at hand. But its long-term impact is just the opposite. When steam-powered plants, jet turbines, car engines, light bulbs, electric motors, air conditioners, and computers were much less efficient than today, they also consumed much less energy. The more efficient they grew, the more of them we built, and the more we used them—and the more energy they consumed overall. Per unit of energy used, the United States produces more than twice as much GDP today as it did in 1950—and total energy consumption in the United States has also risen three-fold. . . . Efficiency fails to curb demand because it lets more people do more, and do it faster—and more/more/faster invariably swamps all the efficiency gains.[15]

Huber and Mills were not the first to conclude that efficiency does not reduce consumption. In 1865, a noted British economist, William Stanley Jevons, published a book that would become his most famous work, *The Coal Question*.[16] Jevons's book was the beginning of what is now known as the field of energy economics. After studying coal consumption patterns in Britain and assuming (wrongly) that his country's coal deposits would soon be exhausted, Jevons concluded, "It is wholly a confusion of ideas to suppose that the economical use of fuels is equivalent to a diminished consumption. The very contrary is the truth."[17] This observation has since come to be known as the Jevons paradox.[18]

In 2003, Vaclav Smil published a magnificent book, *Energy at the Crossroads,* that provides readers with a comprehensive understanding of the history of energy consumption, the problems in forecasting energy use, and the challenges facing any transition away from fossil fuels. When it comes to energy efficiency, Smil—like Huber, Mills, and Jevons—concluded that history is "replete with examples demonstrating that substantial gains in conversion (or material use) efficiencies stimulated increases of fuel and electricity (or additional material) use that were far higher than the savings brought by these innovations."[19]

None of this is offered to imply that efficiency is bad. Efficiency is wonderful. It is an essential part of America's ever-evolving economy. It makes sense to wring more work out of each unit of energy. Energy efficiency conserves capital. It is good for the environment. It is good for rich and poor alike. Efficiency helps reduce the impact of energy price volatility and possible oil-price shocks on consumers. In 2002, two economists at the Congressional Research Service, Marc Labonte and Gail Makinen, wrote a report on this issue concluding that efficiency "has reduced, and can continue to reduce, the effect of these shocks on the overall economy."[20]

While efficiency is laudable, efficiency alone cannot—will not— mean that America uses less energy. Nor will it make the U.S. energy independent. So what about renewable energy and alternative fuels?

12

THE ETHANOL SCAM

Ethanol isn't motor fuel. It's religion. And America is divided into two camps: the believers and the heretics.

So far, it appears that the believers are winning. And their zealotry has created a multi-billion-dollar bushel of subsidies for an energy source that yields no substantive energy benefits for the U.S. That was true back in the 1970s, when the ethanol subsidies began, and it's still true today. The ethanol scam persists because of the religious fervor that pervades any discussion of motor fuel. And just as priests prefer to ignore science when it comes to matters of faith, ethanol boosters prefer to ignore the science that undermines their chosen motor fuel. That religiosity is best summarized by Steffen Schmidt, a political science professor at Iowa State University, who told me that ethanol "is kind of like apple pie. It's almost like a patriotic fuel." There appears, he said, to be a belief "that ethanol is morally better than oil."[1]

Mixing morality and politics can lead to bad policies. But mixing morality and motor fuel makes for a truly lethal cocktail. And few politicians dare look too closely at the ethanol moonshine.

That reluctance has led to some truly lunatic statements. In October 2006, former president Bill Clinton, while in California stumping for Proposition 87 (an alternative-energy initiative that later failed), declared that the initiative would "move California toward energy independence with cleaner fuels, with wind and solar power." He continued,

"There are people who don't believe you can do it. I do. Look at Brazil. Don't you think you can do it if they did it? They run their cars on ethanol."[2] Clinton later provided a sound bite for the pro–Proposition 87 forces in which he declared, "If Brazil can do it, so can California."[3] Clinton didn't bother to mention that all of Brazil's ethanol output—the energy equivalent of about 181,000 barrels of gasoline per day—would be sufficient to last California's motorists for only a little more than four hours per day.[4] (California uses about 1 million barrels of gasoline per day.[5])

The ethanol acolytes include Al Gore and the producers of his documentary, *An Inconvenient Truth*. At the end of the film, in the section about what people can do to help address global warming, this text appears on the screen: "Reduce our dependence on foreign oil, help farmers grow alcohol fuels."[6]

The zealotry over ethanol has led former North Carolina senator and presidential candidate John Edwards to declare that the U.S. should be producing 65 billion gallons of ethanol and other biofuels per year by 2025. In 2007, he published a statement on his campaign Web site saying that his "New Economy Energy Fund" would "develop new methods of producing and using ethanol, including cellulosic ethanol, and offer loan guarantees to new refineries."[7]

The irrational exuberance over ethanol has led investors to dump bushels of cash into the companies that produce ethanol. In June 2006, shares of VeraSun Energy Corp. a North Dakota–based ethanol producer, jumped by more than 30 percent during the first day of trading. The company's initial public offering raised about $420 million, which the company said it would use to build more ethanol-production capacity. Just after it went public, VeraSun's price-to-earnings ratio, a standard method of valuing stocks, was a whopping 200. Just for comparison, Google, one of the hottest companies in America, had a price-to-earnings ratio of 67. That same month, Valero, America's biggest oil refiner, had a price-to-earnings ratio of less than 9. Valero (2006 revenues: $91.8 billion) had refining capacity of about 138 million gallons per day. VeraSun's production capacity was just 630,000 gallons per day. And yet, on a valuation basis, VeraSun was selling for a price that was 22 times as much as Valero's.[8]

Silicon Valley multimillionaire and venture capitalist Vinod Khosla has pumped millions of dollars into ethanol production schemes. And he claims that ethanol will be the "first step" on America's "trajectory to energy independence."[9] But Khosla and his fellow investors are choosing to ignore the many problems faced by the ethanol business. The hard truth about energy in general—and biofuels in particular—is that biofuels can be just as expensive and just as destructive as fossil fuels. In fact, they are almost certainly *more* expensive and *more* destructive.

The myriad problems associated with ethanol prove that the Nobel Prize–winning economist Milton Friedman was right when he famously declared, "There's no such thing as a free lunch." Everything in the energy business comes with a cost. And when it comes to ethanol, the costs are enormous, a fact that has led Nicholas Hollis, the president of the Agribusiness Council, an agriculture trade group, to declare that "ethanol is the largest scam in our nation's history."

Herewith, an itemized invoice of that scam.

BILLIONS IN SUBSIDIES—FOR WHAT, EXACTLY?

If America is "addicted" to oil, then it's equally true that the corn ethanol industry is a world-class junkie when it comes to subsidies. For decades, American politicians have been talking about the need to reduce farm subsidies, and yet, with the ethanol scam, those subsidies, particularly the $0.51-per-gallon tax credit, are thriving.[10]

Making ethanol from corn borders on fiscal insanity. It uses taxpayer money to make subsidized motor fuel from the single most subsidized crop in America. Between 1995 and 2005, federal corn subsidies totaled $51.3 billion. In 2005 alone, according to data compiled by the Environmental Working Group, corn subsidies totaled $9.4 billion.[11] That $9.4 billion is approximately equal to the entire 2006 budget for the U.S. Department of Commerce, a federal agency that has 39,000 employees.[12]

FIGURE 9 U.S. Corn Subsidies, 1995–2005

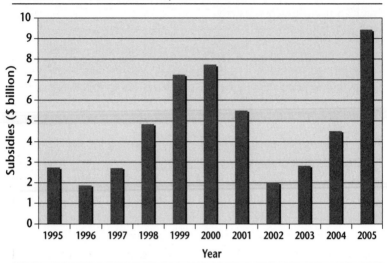

SOURCE: Environmental Working Group.

Corn subsidies dwarf all other agricultural subsidy programs. The $51.3 billion that American taxpayers spent on corn subsidies between 1995 and 2005 was more than twice the amount spent on wheat subsidies, more than twice the amount spent on cotton, almost 4 times the amount spent on soybeans, and 96 times the total subsidies for tobacco in that same period.[13]

But Big Corn isn't satisfied with the subsidies that are paid out to grow the grain. They are also getting massive subsidies to make that grain into fuel. How massive? Well, on February 9, 2007, just a few days after George W. Bush announced his target of 35 billion gallons of renewable and alternative fuel production, the Global Subsidies Initiative—an arm of a Canadian nonprofit group, the International Institute for Sustainable Development—determined that meeting Bush's target by 2017 would require additional subsidies totaling some $118 billion.

The group claims that the $118-billion price tag "would be the minimum subsidy" over the 11-year period it had studied. The authors of the study, Ronald Steenblik of the Global Subsidies Initiative and Doug Koplow of Earth Track, a U.S.-based nonprofit that tracks energy subsidies, said that adding in the tax breaks that the corn distillers are getting from state and local governments and the federal tariffs that are imposed on ethanol from Brazil and other low-cost producers "would likely add tens of billions of dollars of subsidies" to the $118-billion number.[14]

The $0.51-per-gallon tax credit, which is allocated based on the volume of fuel produced, is only part of the subsidy picture. In October 2006, the Global Subsidies Initiative released a 103-page report on ethanol and biodiesel. That report looked at all the government supports for biofuels, including producer tax credits, reductions in state motor fuel taxes, and federal grants, and it found that in 2006, the total subsidies for ethanol ranged from $1.05 per gallon to $1.38 per gallon of ethanol produced. While those are significant numbers, remember that ethanol contains only about two-thirds of the heat value of gasoline. The study concluded that on a gasoline-equivalent basis, the total subsidies for ethanol ranged from $1.42 to $1.87 per gallon.[15] (The spot price for ethanol in early August 2007 was $2.32 per gallon.[16])

The same study found that biofuels subsidies in the U.S. were as much as $7.3 billion per year and that they were likely to rise to as much as $11 billion per year.[17] Thus, the addition of the ongoing subsidies for corn growers—which averaged $4.6 billion per year between 1995 and 2005—to the ethanol production subsidies means that American taxpayers may soon be paying nearly $16 billion per year to subsidize the production of a motor fuel that will do little, if anything, to reduce America's overall oil imports. Just for the sake of comparison, $16 billion is about three times as much as the federal government spends per year on WIC, short for the Special Supplemental Nutrition Program for Women, Infants and Children, a program that provides health care and nutrition assistance to low-income mothers and children under the age of five.[18]

The biofuel subsidies are allowed to proliferate for a simple reason: Their backers continually discuss the need for ethanol and biodiesel in terms of the war in the Middle East and reducing the amount of foreign oil imported. The problem with those arguments is that the ethanol backers cannot, or will not, ever say exactly what level of imports is acceptable. That is, if it costs, say, $11 billion to reduce foreign oil imports by 1 percent, is that a good bargain for the taxpayers? Is the goal to lower oil imports from 60 percent to 53 percent, or should it be 49.32 percent?

The ethanol boosters don't discuss the optimum level of oil imports for a simple reason: They are not really interested in reducing imports. Instead, their main interest is in ensuring that the flow of taxpayer subsidies will continue unabated. But those subsidies cannot make America energy independent. That was the conclusion of a 1997 report by the Government Accountability Office, which concluded that the tax incentives given to the ethanol producers "do not significantly reduce petroleum imports. Therefore, the tax incentives do not significantly contribute to U.S. energy independence."[19]

Tax incentives won't result in energy independence, but they will mean more profits for the world's largest agribusiness outfit, Archer Daniels Midland Co. (ADM).

THE ADM IMPERATIVE AND THE "AGRICULTURAL EQUIVALENT OF HOLY WATER"

In 2006, agribusiness giant Archer Daniels Midland Co. announced that it had hired Patricia Woertz, a former Chevron executive, to be its new CEO. When Woertz got to ADM, she took the helm of an outfit that controlled 29 percent of the U.S. ethanol market. According to *Barron's,* no other player in the market at that time controlled more than 5 percent of the ethanol trade.[20]

Woertz's old company, Chevron, has about 1.5 million barrels of oil-refining capacity in the U.S. That means that Chevron, the second-largest integrated oil company in the U.S., controls about 7.5 percent of the domestic market for refined products.[21] The biggest U.S. refiner is Houston-based ConocoPhillips, which has about 13 percent of America's refining capacity.[22] That means that ADM's share of the U.S. ethanol market is about two times as large as the market share in petroleum products enjoyed by any of the big oil refiners. And ADM's share of the ethanol business is nearly six times as large as that of any other ethanol producer. That gives ADM a significant amount of market power in the ethanol business. Shortly after being hired by ADM, Woertz acknowledged her new company's clout. She declared that biofuels and ethanol are "an area where ADM can have some competitive advantage."[23]

Yes, indeed. And if any company in America has demonstrated a willingness to exploit its "competitive advantage," it's ADM, an admitted price fixer.[24]

In 1996 ADM was indicted for manipulating the price of lysine and citric acid. (Lysine is an amino acid that is used as an additive in feed given to poultry, swine, and other livestock. Citric acid is a common food additive.) To settle the case, which involved two felony charges against ADM, the company agreed to pay a criminal fine of $100 million. At that time it was the largest criminal antitrust fine ever levied in America.[25] In fact, the fine was seven times higher than any fine ever collected by the U.S. Department of Justice. In announcing the settlement of the case, Assistant U.S. Attorney General Joel Klein said that "greed, simple greed, replaced any sense of corporate decency or integrity."[26] In 1999, three ADM executives were convicted on criminal charges of price fixing and were given prison sentences.[27] One of those executives was Mick Andreas, the son of Dwayne Andreas, the longtime chairman of ADM. Remarkably, the elder Andreas avoided being indicted on price-fixing charges. Further, according to *Rats in the Grain,* an excellent book on the ADM price-fixing scandal written by James B. Lieber, Dwayne Andreas had systematically destroyed a significant number of documents shortly after the Department of Justice began its investigation into ADM's shenanigans.[28] Lieber also reports that ADM executives employed prostitutes in an effort to get information from some of their rivals.[29]

Perhaps more significant than the destruction of documents was a statement attributed to Dwayne Andreas that was ADM's unofficial motto when it came to fixing prices: "The competitor is our friend and the customer is our enemy."[30]

After its investigation was completed, the Department of Justice slapped ADM with a $100-million fine. The company also paid some $90 million to settle the civil cases. But the $190-million tariff was a relatively minor cost for ADM. As Lieber reports in *Rats in the Grain*, the company made some $300 million in profits by manipulating the prices of lysine and citric acid.[31]

The amorality at ADM stunned some of the judges who heard appeals on the price-fixing case. In mid-2000, judges at the Seventh Circuit Court of Appeals wrote that the facts in the case "reflect an inexplicable lack of business ethics and an atmosphere of lawlessness."[32]

The lysine and citric acid convictions led to another fine. In 2004, the company paid a $400-million fine to settle a civil suit alleging it had rigged prices on high-fructose corn syrup, an ingredient used in hundreds of consumer foods.[33]

And yet, the massive fines that ADM paid are, it appears, just a cost of doing business for the company that calls itself the "supermarket to the world." Another of ADM's costs of doing business: huge political campaign contributions. ADM's history is replete with examples of how Dwayne Andreas and his family have used political campaign contributions to help keep the federal subsidies flowing on everything from corn production, to corn sweeteners, to ethanol. And those subsidies have helped make ADM into a powerhouse.

In 1995, Andreas flatly declared, "There isn't one grain of anything in the world that is sold in a free market. Not one! . . . People who are not in the Midwest do not understand that this is a socialist country." Andreas's devotion to socialism in American agriculture goes back to the Nixon era. In 1972, he delivered an envelope to Nixon's office that contained $100,000 in $100 bills. The money was stashed in a White House safe for several months. But as the Watergate disaster began to unfold, Nixon decided to return the cash to Andreas. Nevertheless, a bank account that was used by the Watergate burglars was later found to contain $25,000 that was traced back to Andreas. For years, Andreas's main proxy in the U.S. Senate was Bob Dole, the Kansas Republican, who received huge campaign contributions from ADM. In 1980, Dole sponsored the bill to impose tariffs on foreign ethanol. In 1990 and 1992, Dole pushed proethanol amendments onto various pieces of Senate legislation. And by the mid-1990s, ADM was producing 60 percent of all the corn ethanol in America.[34]

To keep the federal subsidies flowing, ADM became an ATM for favored politicians. Between 1989 and 2006, ADM contributed nearly $7.9 million to politicians from both parties, an amount that made it the 85th largest political donor in the U.S. over that time period.[35] The list of ADM's recipients include some of the Senate's biggest boosters of ethanol. According to the Center for Responsive Politics, Senator Dick Durbin, an Illinois Democrat, received the most from ADM ($57,350) between 1989 and 2005. Durbin has sponsored numerous pieces of legislation to increase ethanol consumption. Another big recipient of ADM's cash: Iowa's senator, Tom Harkin, another Democrat and a big ethanol proponent. During that same time span, Harkin got $42,000 from ADM.[36]

ADM has given large amounts of money to the last two Republican Speakers of the House, Newt Gingrich and Dennis Hastert. In 1995, Dan Carney reported in *Mother Jones* magazine that Andreas gave "at least $70,000" to Gingrich's political action committee, GOPAC.[37] Hastert took in $38,500 from ADM between 1989 and 2006, making him the fifth-largest individual recipient of the company's money during that time period.[38]

ADM and Andreas have also given out lots of "soft money" (which is not constrained by federal campaign laws) to various political causes. In 2001, ADM gave $100,000 to George W. Bush's inaugural committee.[39] During the 2002 election cycle alone, ADM handed out $1.7 million in soft money, the majority of which went to Republicans. That giving ranks ADM as the 26th largest soft money donor during that time period.[40] Between 1990 and 1998, Andreas and his wife, Inez, gave $673,000 in soft money, most of it to Democrats.[41]

What did all that money buy? It purchased the continuation of the socialism that Dwayne Andreas and his family have grown to love. In 1995, James Bovard of the Cato Institute published a study estimating that every $1 that ADM earns from ethanol costs taxpayers $30. Bovard's report, which contains 138 footnotes, includes one of the single best paragraphs ever written about the ethanol scam and its effect on the elected class:

> Ethanol has become a magic obeisance button for politicians. Simply mention the word and politicians grovel like trained dogs, competing to heap the most praise on ethanol and its well-connected producers. Regardless of how uncompetitive the product may be, politicians have for years talked about ethanol as if it were the agricultural equivalent of holy water.[42]

Bovard went on to estimate that "at least 43 percent of ADM's annual profits are from products heavily subsidized or protected by the U.S. government."[43] And ADM has been extraordinarily effective at protecting those subsidies and the profits they generate. The best example is the 1998 effort by the powerful chairman of the House Ways and Means Committee, Bill Archer, a Texas Republican, to kill the ethanol subsidy altogether. Archer had all the votes that he needed in the House to eliminate the tax credits for ethanol. (At that time, the tax credit was $0.54 per gallon.) But it appears that ADM's cash changed Gingrich's mind. Archer's effort to kill the ethanol subsidy was derailed when Gingrich stacked a House-Senate conference committee with proethanol members. Shortly after the vote, the *Washington Post* reported that "an angry Archer said Gingrich had given in to pressure from farm-state members worried about their reelection campaigns."[44]

Given ADM's history as a price-fixing–influence-buying company that has shown little regard for antitrust laws and fair competition, why are the environmentalists and farmers and labor unions so proethanol? In particular, why are they rushing to hand ADM—one of the most notorious price gougers in the history of American business—a dominant share of the ethanol market? Do they really believe that ADM is going to be more benevolent than Exxon Mobil or Chevron?

Those questions are not easily answered. But it's clear that the ethanol colossus created by ADM has developed a momentum of its own. In the three decades that have passed since Dwayne Andreas's cash was stashed in Richard Nixon's White House safe, ethanol has become an electoral juggernaut—one that forces presidential candidates to genuflect before it lest they get ambushed in Iowa.

THE IOWA IMPERATIVE AND
JOHN MCCAIN'S "HIGHWAY ROBBERY"

John McCain relishes his image as a truth teller. Unfortunately, his bus—dubbed the Straight Talk Express—got hijacked in Iowa.

For years, McCain has been among America's most ardent critics of ethanol. Back in 1999, during a Des Moines debate, he declared, "Ethanol is not worth it. It does not help the consumer. Those ethanol subsidies should be phased out."[45] In 2002, the Republican senator from Arizona declared that ethanol is a "giveaway to special interests in corn-growing states at the expense of the rest of the country."[46]

In 2003, McCain issued a scathing press release regarding a then-pending energy bill. He declared that ethanol wouldn't exist if Congress didn't "create an artificial market for it." He went on, saying that ethanol "does nothing to reduce fuel consumption, nothing to increase our energy independence, nothing to improve air quality. Let me repeat for emphasis; ethanol does nothing to reduce fuel consumption, nothing to increase energy independence and nothing to improve air quality."[47] After a few more sentences detailing his opposition, McCain said, "Plain and simple, the ethanol program is highway robbery perpetrated on the American public by Congress."[48]

In 2005, McCain voted against the big energy bill that passed Congress, because its ethanol mandates would "result in higher gasoline costs for states, like Arizona, that do not have an abundant in-state supply" of ethanol.[49]

But all of McCain's persecution of the ethanol faithful happened before he decided to make a serious run for the White House. And his conversion, like that of Saint Paul, has transformed him from an ethanol atheist into an ardent evangelist. McCain's now among a pew-full of presidential aspirants who worship at the altar of ethyl alcohol and sing from the same hymnal. In an August 2006 speech in Grinnell, Iowa, McCain revealed that he has become a corn convert, an ethanol evangelical.

During that speech, he declared that ethanol is "a vital alternative energy source not only because of our dependency on foreign oil but its greenhouse gas reduction effects."[50] In February 2007, the new McCain declared that America needs to be energy independent and "obviously ethanol is a big part of that equation."[51]

Call it the Iowa imperative. Any candidate who wants to win the White House must have a good showing in the very first presidential primary—the Iowa caucuses—which were held January 3, 2008.[52] The imperative can be explained by looking at the numbers: By early 2007, Iowa was producing about one-third of all of the ethanol in the U.S.[53] Since 2002, the amount of Iowa corn going into ethanol production has tripled. By mid-2007, the state had 24 producing ethanol plants, and another 22 were either planned or under construction.[54] About 27,000 jobs were directly related to ethanol production, and another 20,000 people had jobs—according to IowaCorn.org—that were "affected" by ethanol.[55]

If you are a presidential candidate, supporting corn ethanol shows that you believe in the redeeming power of subsidies. And subsidies are a major factor in Iowa's economy. Iowa gets more agricultural subsidies than all but one other state (Texas). Between 1995 and 2005, U.S. taxpayers doled out some $14.7 billion in agricultural subsidies to Iowa farmers.[56] In 2005 alone, those subsidies totaled $2.24 billion, or about $757 for every resident of the state.[57] Further, of that $2.24 billion, about $1.8 billion came from corn subsidies.[58] Thus, in 2005, corn subsidies provided about $608 for every Iowan.

Add the massive subsidies to the power of the ethanol lobby and you get a political dynamic that has perverted America's energy policy and presidential selection system. The Iowa imperative allows the economic needs of a state with less than 1 percent of the U.S. population—or to be more precise, the economic desires of a tiny subgroup of that 1 percent, that is, Iowa's corn farmers and ethanol producers—to dictate the terms of a debate that affects every American taxpayer.[59]

The Iowa imperative has long been the key factor for presidential contenders. Prior to running for president in the 2000 campaign, former New Jersey senator Bill Bradley, a Democrat, was a leading critic of ethanol. He once called the ethanol subsidy a "billion-dollar tax break [that] is nothing more than a gift to a single, politically connected industry."[60] He also said the ethanol subsidy meant higher gasoline prices for his New Jersey constituents. But once Bradley began eyeing the White House, he became an ethanol convert. In early 2000, while campaigning in Iowa, Bradley told voters that he had changed his mind on ethanol, stating that the U.S. had to "help our family farmers get a bigger chunk of the food dollar."[61]

The same flipping and flopping that afflicted McCain and Bradley has also savaged Hillary Clinton, who as a U.S. senator from New York voted against ethanol some 17 times. In 2002, she signed a letter saying that the ethanol subsidies were "equivalent to a new tax" on gasoline and that there is "no sound public policy reason for mandating the use of ethanol." In 2005, Clinton opposed the final version of an energy bill that provided grants to cellulosic ethanol producers and opposed an amendment that was to require refiners to increase their use of renewable fuels.

But in early 2007, during a visit to Des Moines, Clinton said that the U.S. needs to work on "limiting our dependence on foreign oil. And we have a perfect example right here in Iowa about how it can work with all of the ethanol that's being produced here."[62]

But it's not enough to simply support the production of ethanol. Domestic producers must also be insulated from foreign competition. In 2006, Barack Obama, the Illinois Democrat, along with four other farm-state senators, sent a letter to President Bush asking him to ignore calls to reduce tariffs on Brazilian sugarcane-based ethanol. Lowering the tariff, they said, would make the U.S. dependent on foreign ethanol. "Our focus must be on building energy security through domestically produced renewable fuels," they wrote. Perhaps it's just a coincidence, but during his first year in office, Obama twice used corporate jets belonging to the Illinois-based agribusiness giant ADM, America's biggest ethanol producer.[63]

The Iowa imperative shows up in the polls: A January 2007 poll found that 92 percent of Iowa voters believed that ethanol was important to the state's economic future.[64] As Iowa State University's Steffen Schmidt put it, there is "sort of a halo over all the biofuels. That's why when politicians come to Iowa, they have to say ethanol is great."[65]

And they do. In 2008, not one of the leading presidential candidates has dared say anything negative about ethanol. Amid the deafening silence from the candidates, the "highway robbery" that McCain talked about back in 2003 continues apace. And Iowans continue reaping the rich subsidies that flow during the silence.

FOOD OR FUEL?

In January 2007, three U.S. senators—two Democrats, Tom Harkin of Iowa and Barack Obama of Illinois, along with Indiana Republican Richard Lugar—introduced a bill that would promote the use of ethanol. It also mandates the use of more biodiesel and creates tax credits for the production of cellulosic ethanol. The three senators called their bill the American Fuels Act of 2007.

The most amazing part of the press release trumpeting the legislation is its fourth paragraph, in which Lugar declares that "U.S. policies should be targeted to replace hydrocarbons with carbohydrates."[66] Nor is Lugar the only one promoting a plan that would turn food into fuel. Lugar's fellow ethanol booster, James Woolsey, declared in a March 2007 speech to the Virginia Soybean, Corn, and Grain Association, "We must move from a hydrocarbon-based society to a carbohydrate-based society."[67]

Let's consider Lugar's and Woolsey's proposal for a moment. In the 18th and 19th centuries, the U.S. economy was primarily based on carbohydrates. For most people, horses were the main mode of transportation. They were also a primary work source for plowing and planting. Aside from coal, which was used by the railroads and in some factories, the U.S. economy depended largely on the ability of draft animals to turn forage into usable work. America's farmers were solely focused on producing food and fiber. And while the U.S. was moderately prosperous, it was not a world leader.

Oil changed all that. After the discovery of vast quantities of oil in Pennsylvania, Texas, Oklahoma, California, and other locales, America was able to create a modern transportation system, with cars, buses, and airplanes. That oil helped the U.S. become a dominant military power. Humans were freed from the limitations of the carbohydrate economy, which was constrained by the amount of arable land. Today's hydrocarbon-based economy has made America's arable land far more productive by allowing farmers to cultivate their land more intensively and concentrate on growing food rather than fodder.

In short, hydrocarbons were a critical ingredient in making the U.S. into one of the most prosperous countries on earth.

But for some reason, Lugar, Woolsey, and their cohorts long for a return to the old days, one in which the vagaries of rainfall and periodic drought will determine whether America's transportation system has the fuel that it needs. A severe, years-long drought in Iowa or Illinois or other key corn-producing states could devastate America's ethanol industry, and the U.S. would have no choice but to make up for any shortfall in ethanol production by importing more oil. In addition, it's reasonable to assume—given the ethanol industry's decades-long addiction to subsidies—that federal taxpayers will be called upon to prop up the ethanol producers in the event of such a drought.

While the effects of weather-related calamities are rarely discussed, Lugar, Woolsey, and the other neo-Luddite energy isolationists are purposely ignoring an even more fundamental question: Should America's farms be producing food or fuel?

In September 2006, Lester Brown, the president of the Earth Policy Institute, a group that promotes "an environmentally sustainable economy," wrote in a *Washington Post* opinion piece that the amount of grain needed to make enough ethanol to fill a 25-gallon SUV tank "would feed one person for a full year." Brown said the ongoing ethanol boom in the U.S. was "setting the stage for an epic competition. In a narrow sense, it is one between the world's supermarkets and its service stations." He continued, "It is a battle between the world's 800 million automobile owners, who want to maintain their mobility, and the world's 2 billion poorest people, who simply want to survive."[68]

Using food to make fuel bothers many food analysts, and their political leanings—liberal or conservative—don't seem to matter. Dennis Avery, director of global food issues at the Hudson Institute, a conservative think tank in Washington, D.C., has concerns that are remarkably similar to Brown's. A few days after Brown's piece appeared in the *Post*, Avery published a paper for another conservative think tank, the Competitive Enterprise Institute, which showed that ethanol simply cannot provide enough motor fuel to make a significant difference in America's fuel consumption. And like Brown, Avery laid bare the essential question: Food or fuel?

"The real conflict over cropland in the 21st century," wrote Avery, "will set people's desire for biofuels against their altruistic desire that all the children on the planet be well-nourished." He continued, "The world's cropland resources seem totally inadequate to the vast size of the energy challenge. We would effectively be burning food as auto fuel in a world that is not fully well-fed now, and whose food demand will more than double in the next 40 years." Avery said that even if the U.S. adopted biofuels as the solution for imported crude oil, "it would take more than 546 million acres of U.S. farmland to replace all of our current gasoline use with corn ethanol." That's a huge area, especially as the total amount of American cropland covers about 440 million acres.[69]

While farmers and Big Agriculture prefer to cast the use of ethanol and biofuels in terms of national security, foreign oil, and rural development, the larger issue is a moral one: Are we going to use our precious farmland to grow food, or are we going to subsidize the growth of an industry that turns food into a commodity—motor fuel—of which we already have an adequate supply?

The morality of biofuel production even caught the eye of Cuban dictator Fidel Castro. In March 2007, shortly after George W. Bush met with the leaders of the big Detroit automakers at the White House to promote the use of ethanol, Castro wrote a letter that was published by the state-owned newspaper, *Granma*. Castro declared that "converting food into fuel" was a "sinister idea" and that the expanded use of biofuels would cause hunger among the people in the poorer countries of the world.[70]

In May 2007, researchers from Iowa State University's Center for Agricultural and Rural Development released a report that looked at how ethanol production—which consumed 20 percent of America's corn crop in 2006—is affecting overall food prices.[71] They found that ethanol's voracious appetite for grain has resulted in higher prices on a panoply of foods, including cheese, ice cream, eggs, poultry, pork, cereal, sugar, and beef. They found that between July 2006 and May 2007, the food bill for every American increased by about $47 due to surging prices for corn and the associated price increases for other grains like soybeans and wheat. The Iowa State researchers also concluded that in aggregate, American consumers face a "total cost of ethanol of about $14 billion."[72]

Let's put that $14 billion in perspective.

In 2006, the U.S. produced 5 billion gallons of corn ethanol. That means that Americans are effectively paying a new tax (in the form of higher food costs) of $2.80 for each gallon of ethanol being produced. Considering the amount of oil that is actually replaced by corn ethanol, the food tax is about $3.72 per gallon of gasoline displaced. Here's why: The lower heat value of ethanol requires consumers to buy about 1.33 gallons of ethanol to get the equivalent of a gallon of gasoline. ($2.80 × 1.33 = $3.72). And that food tax of $3.72 per gallon of gasoline displaced does *not* count any of the multi-billion-dollar federal subsidies for corn production or the $0.51-per-gallon subsidy given to the companies that distill ethanol from corn.

In short, American taxpayers are being taxed three different ways in order to produce corn ethanol: (1) the billions in subsidies for growing corn; (2) the billions in subsidies for turning that corn into ethanol; and (3) the billions of dollars in costs that come from higher food prices. And it's not just Americans who are being fleeced. Thanks to ethanol's voracious appetite for grain, the Iowa State researchers determined that "the rest of the world's consumers would also see higher food prices."[73]

This new triple-tax regime comes courtesy of ADM, Big Agriculture, and a corrupt Congress. And that tax will ultimately take food out of the mouths of human beings.

CORN ETHANOL'S ENRON ACCOUNTING

In 2004, the American Coalition for Ethanol published a paper to rebut the claims of scientists who were claiming that ethanol uses more energy than it produces. The report cited several proethanol studies that had claimed ethanol was a net energy gainer. It went on, saying that "ethanol lessens America's reliance on foreign countries for oil" and "keeps our dollars home." It then concluded that "when looking at all of the facts, it makes counting Btus seem rather silly."[74]

Actually, it's not silly at all. The issue of whether ethanol is worth doing from an energy standpoint is absolutely critical. It goes to the very heart of the rationale behind the ethanol craze. No rational investor would put money into a stock or other investment that he or she knew was going to lose money. The same is true of energy. The problem, as always, comes during the accounting. The top bosses at Enron, like Jeff Skilling, Ken Lay, and Andrew Fastow, were experts at perverting their company's accounting. Similar problems are afoot in the accounting to track the net energy gains (or losses) associated with corn ethanol production. And just as in financial accounting, the entire game hinges on who does the accounting and how expenses are accounted for.

The process of determining a given process's energy profitability is called the EROEI, short for energy returned on energy invested. It's also called the energy balance. Enron excelled at hiding its debts and expenses and, in doing so, made the company look more profitable (in dollar terms). Corn ethanol producers excel at hiding their energy expenses and, in doing so, make their process look more profitable (in Btu terms).

For instance, in a widely cited 2002 study, three scientists from the U.S. Department of Agriculture (USDA) produced a report claiming that ethanol produces about 34 percent more energy than it requires to produce. Put another way, for every 1 Btu invested, an investor gets 1.34 Btus in return.[75] But the details—as always—are in the fine print. The scientists are able to achieve that 34 percent net gain only by including "coproducts energy credits," that is, by adding in the energy value of the by-products that are created during the ethanol production process. Among the most important of these by-products is DDG, short for dried distiller's grain.[76] That's the meal left over from the corn after it has been crushed and fermented. Ethanol advocates insist the coproduct credits are reasonable because the DDG can be used as cattle feed. That may be true, but you can't run your car on DDG. Therefore, eliminating the coproduct credit provides a clearer accounting statement when it comes to corn ethanol's energy balance.

What happens if the coproduct credit is eliminated? That's where it gets interesting. The coproduct credit is the difference between energy profitability and just above break-even. Without that credit, the total energy profits from the corn ethanol plummet to about 9 percent. Thus, for every 1 Btu invested, an investor gets 1.09 Btus in return. As explained by Robert Rapier—a clever chemical engineer (and native of Oklahoma) who writes the R-Squared Energy Blog, one of the best blogs on energy and the energy business—the USDA scientists used "a completely illegitimate accounting trick to exaggerate" the energy profits of the ethanol production process.[77]

The paltry energy profitability of corn ethanol has been corroborated by numerous studies. For instance, in January 2007, researchers at Oregon State University looked at biofuels and came up with a similar energy profitability for corn ethanol. They concluded that corn ethanol's net energy gain was "just 20.3% of the energy contained in the fuel."[78] That is, for 1 Btu invested, an investor gets 1.2 Btus in return. But even with that slight gain in energy, the Oregon State report concluded that corn-based ethanol was "significantly more costly than gasoline."[79]

Ethanol has loud and persistent critics in the scientific community, and in their accounting, corn ethanol yields only energy losses. The most famous of these critics are David Pimentel, a professor of ecology at Cornell University who has been studying grain alcohol for two decades, and Tad Patzek, an engineering professor at the University of California at Berkeley. In 2005, they cowrote a report claiming that corn ethanol production results in a net energy loss of 29 percent; that is, for 1 Btu invested, an investor gets 0.71 Btus in return. In addition to their findings on corn, they determined that cellulosic ethanol created from switchgrass results in a 50 percent net loss, and ethanol from wood biomass has a 57 percent net loss of energy. The best yield comes from soybeans, but they, too, are a net loser, according to Pimentel and Patzek, who claim that biodiesel produced from soybeans consumes 27 percent more fossil energy than it contains.[80] (Patzek and Pimentel have produced a number of other studies on corn ethanol production, none of them flattering.)

Of course, producing motor fuel requires huge energy inputs as well. From drilling the well, to the pipelines, to the refinery, oil companies consume enormous quantities of fossil fuel in order to produce diesel, gasoline, and jet fuel. But these processes are far more efficient than those used to produce corn ethanol because the feedstock (crude oil) has a much higher energy density than grain. The energy accounting for gasoline production shows that it yields energy profits of about 600 to 700 percent. Put another way, for 1 Btu invested in crude oil and gasoline production, an investor gets 6 or 7 Btus back.[81] (Other studies have shown returns of up to 2,000 percent for gasoline.[82]) That high rate of return—at least 22 times as much as the energy profits from corn ethanol—helps explain why oil-based fuels have been used so profitably, for so long. They have very high energy content, are fairly light, and are easily managed and transported.

This discussion of energy profits leads directly to another critical question: What are corn ethanol's total carbon dioxide emissions? Given its dubious energy profitability, it makes sense to assume that corn ethanol's carbon emissions would be close to, or equal to, the emissions from traditional oil-based fuels. And in fact, several government studies have found exactly that. In 1997, the Government Accountability Office determined:

> The global-warming effects of using ethanol are likely to be no better than, and could be worse than, those of using conventional gasoline. Furthermore, even if current ethanol use were to contribute to lowered greenhouse gas emissions, ethanol is such a small part of total U.S. fuel use that global environmental quality should not be significantly affected if ethanol use were discontinued.[83]

That same 1997 study said that "the global-warming picture may be worsened by using ethanol." The Government Accountability Office found that the greenhouse gases released during the ethanol fuel cycle contain "relatively more nitrous oxide and other potent greenhouse gases. In contrast, the greenhouse gases released during the conventional gasoline fuel cycle contain relatively more of the less potent type, namely, carbon dioxide."[84]

Other government studies have come to similar conclusions. A 2005 study by the Energy Information Administration found that corn ethanol made by distilleries that use coal as a primary fuel source produces fuel that yields just 7 percent fewer greenhouse gas emissions than standard gasoline.[85] In an October 2006 report on corn ethanol, the Congressional Research Service determined that the "benefits in terms of greenhouse gas emissions reductions is [sic] limited."[86]

Scientists outside government are reaching similar conclusions. In January 2006, a study published in *Science* magazine found that "corn ethanol technologies are much less petroleum-intensive than gasoline but have greenhouse gas emissions similar to those of gasoline."[87] In early 2007, two fellows at the Cato Institute wrote that gasoline mixed with 10 percent ethanol cut greenhouse gas emissions from 0 to 5 percent, while E85—motor fuel containing 85 percent ethanol and 15 percent gasoline—could cut emissions by 12 percent. But they said there are far cheaper ways to reduce greenhouse gas emissions.[88]

Yet another study, released in June 2007 by two Colorado scientists—Jan F. Kreider, an engineering professor at the University of Colorado, and Peter S. Curtiss, a Boulder-based engineering consultant—determined that not only does corn ethanol produce little, if any, new energy, but its carbon dioxide emissions are also worse than those of conventional gasoline and diesel fuel. Their peer-reviewed paper, which was presented at a conference sponsored by the American Society of Mechanical Engineers in June 2007, concludes that during the entire life cycle of ethanol, carbon dioxide emissions are "about 50% larger for ethanols than for traditional fossil fuels; such fuels are not the answer to global warming, they make it worse."[89]

So what's the bottom line? Well, the Congressional Research Service (CRS) neatly summarized the energy profitability question in its October 2006 report. It wrote that ethanol has the potential to "significantly displace *petroleum* demand" (italics in original).

But here's the essential part of that report: The CRS researchers determined that "with only a slight net energy benefit from the use of corn-based ethanol, transportation energy demand is essentially transferred from one fossil fuel (petroleum) to another (natural gas and/or coal)."[90]

In other words, corn ethanol production is just a shell game; it merely changes one form of energy into another. To put it another way, corn ethanol takes expensive Btus from the left hand and passes the same number of Btus (in a slightly different form) to the right hand, and as it does so, Big Agriculture and Big Corn get to take a Big Subsidy. And all along that chain of Btu legerdemain, no real energy profits are ever made.

The point here is painfully obvious. Corn ethanol supporters like to use Enron-style accounting methods and want to marginalize the "silly" counting of Btus for a simple reason: An impartial accounting process shows that ethanol provides marginal, if any, energy profits. Can you imagine any company in America telling its shareholders that it didn't want to be bothered with "silly" exercises like, say, counting the money in the till? And yet, the most fundamental questions about corn ethanol—its energy profitability and its carbon dioxide emissions—continue to be obscured as the ethanol boosters prefer to prattle on about "energy security" and the evils of Saudi Arabia and OPEC.

When not talking about Saudi Arabia, the energy isolationists and ethanol boosters like to obsess about another country: Brazil.

ETHANOL MADNESS WITH A BRAZILIAN BEAT

Any discussion of ethanol inevitably leads to a discussion of the wonders of Brazil's ethanol industry, which turns sugarcane into motor fuel. In 2007, George W. Bush traveled to São Paolo to discuss, among other things, the prospects for increasing the amount of Brazilian ethanol that can be provided to the U.S. market. During a March 9 speech in São Paolo, Bush said that "as we diversify away from the use of gasoline by using ethanol we're really diversifying away from oil." He went on, saying that "we all feel incumbent [sic] to be good stewards of the environment. It just so happens that ethanol and biodiesel will help improve the quality of the environment in our respective countries. And so I'm very much in favor of promoting the technologies that will enable ethanol and biodiesel to remain competitive, and therefore, affordable."[91] During his visit, Bush declared that he was "very upbeat about the potential of biofuel and ethanol." He also signed an agreement that calls for Brazil and the U.S. to increase investments in biofuels and work on technology exchanges.[92]

The same day Bush was in Brazil, the *Wall Street Journal* ran a front-page story that looked at the booming ethanol trade in Latin America. It said that encouraging ethanol output was a "top energy priority" for

Bush and that he was "encouraging alternative sources such as ethanol."[93] A few days before Bush got to Brazil, *The Economist* declared that
"ethanol diplomacy" would "be a focus of Mr. Bush's Latin American
tour."[94] Several weeks earlier, the *Washington Post* had reported that the
U.S. and Brazil were talking about an ethanol partnership that would
"diminish the regional influence of oil-rich Venezuela."[95]

Politicos of all stripes view Brazil as a biofuel paradise. In late 2005,
in a speech to the National Press Club, Pennsylvania governor Edward
Rendell said, "No longer is investing in alternative fuels a fringe idea.
. . . Brazil is perhaps the world's greatest success story. Due to 30 years
of hard work, research and investment, Brazil will not need one drop
of imported oil this time next year. If anyone suggests to you that these
ideas aren't ready for prime time and cost too much, they are living in
the past."[96]

Venture capitalist Vinod Khosla and former Senate minority leader
Tom Daschle have touted Brazil's "energy independence miracle." In a
May 2006 opinion piece in the *New York Times,* they said that ethanol
"could set America free from its dependence on foreign oil" and that
Brazil proves that "an aggressive strategy of investing in petroleum
substitutes like ethanol can end dependence on imported oil." They
went on to tout cellulosic ethanol, saying that ethanol "produced from
perennial energy crops like switch grass can slash our carbon dioxide
emissions."[97]

Amid all the misbegotten propaganda, something has been lost:
perspective. Yes, Brazilian ethanol may be cheaper than domestically
produced corn ethanol. And yes, the $0.54-per-gallon tariff on foreign
ethanol is perhaps the most egregious example of protectionism and
provincialism in America's convoluted energy policy. But none of the
politicos or journalists who are blathering about Brazil have bothered
to count the number of Btus actually involved in the ethanol trade.
Nor have they bothered to examine how ethanol fits into the larger
picture of energy consumption in Brazil.

A big reality check is in order.

First and foremost, Brazil is not the epicenter of ethanol production; the U.S. is. In both 2005 and 2006, the U.S. produced more

ethanol than Brazil.[98] Further, U.S. production is growing by 20 percent per year. Brazil's ethanol production in 2005 and 2006 was essentially flat.[99] In 2005, ethanol supplied about 3 percent of the volume of gasoline used in the U.S. In Brazil, it supplied about 40 percent of the gasoline consumption.[100] Those numbers reflect the huge disparities in both population and vehicles. Brazil has about 190 million people, but just 23 million automobiles.[101] The U.S. has 110 million more people than Brazil and more than 10 times as many motor vehicles. Those numbers are reflected in the fact that the U.S. has more motor vehicles than it does registered drivers. According to Cambridge Energy Research Associates, the U.S. has 1,148 registered personal vehicles for every 1,000 licensed drivers. Brazil, by contrast, has just 137 personal vehicles for every 1,000 licensed drivers. (For the sake of comparison, India has 11 cars per 1,000 drivers, and China has 9 cars per 1,000 eligible drivers.[102])

Brazil has a huge advantage in ethanol production because sugarcane is a far better feedstock than corn. While corn ethanol provides very little, if any, energy profits, ethanol produced from sugarcane produces about 8 times more energy than is required to produce the fuel.[103] That is, for every 1 Btu invested, an investor gets about 8 Btus of energy profits. Brazil's tropical climate makes it perfect for sugarcane production. Brazil produces about twice as much sugarcane as India and about 16 times as much as the U.S.[104] Petrobras, Brazil's national oil company, claims it can produce twice as much ethanol per acre as U.S. corn farmers and do so at half the cost.[105]

Brazil has plenty of cheap labor to harvest that sugarcane.[106] Cutting sugarcane is dangerous, backbreaking work. In an article published on the same day that Bush was in São Paolo talking about ethanol, London's *The Guardian* newspaper ran a story quoting human rights activists who said that the men who harvest sugarcane for ethanol production "are effectively slaves" and that Brazil's ethanol industry is "a shadowy world of middle men and human rights abuses." It cited mortality figures provided by a Catholic nun, Sister Ines Facioli, who runs a support network in a small town about 200 miles west of São Paolo. She claimed that between 2004 and 2006, 17 cane workers had died due to overwork or exhaustion. One laborer, Pedro Castro, told *The Guardian*'s Tom Phillips, that the hot climate, combined with the heavy protective clothing needed to protect his body from the sharp machete blades used to cut the cane, was like working "inside a bread oven."[107]

For the work, the average cane worker gets paid about $1 for every ton of sugarcane cut. The workers often labor in 12-hour shifts. Their housing, according to Phillips's article, consists of "squalid, overcrowded 'guest houses' rented to them at extortionate prices by unscrupulous landlords." The average cane cutter makes less than $200 per month.[108] And some, it appears, make nothing at all.

In July 2007, the Brazilian government freed 1,100 laborers who were found working in horrendous conditions on a sugarcane plantation in the northeastern state of Para. A story by the Associated Press said that the workers were forced to work 13-hour days and that they had no choice but to pay "exorbitant prices for food and medicine." It then cited a source in Brazil's labor ministry who claimed that many of the workers were "sick from spoiled food or unsafe water, slept in cramped quarters on hammocks and did not have proper sanitation facilities." The government-backed raid of the plantation lasted three days. The plantation in question is owned by Para Pastoril e Agricola SA, which produces about 13 million gallons of ethanol per year. The workers were caught up in a situation known as debt slavery, in which poor workers are taken to remote farms, where they rack up large debts to the plantation owners, who force the workers to pay high prices for everything from food to transportation.

According to Land Pastoral, a group affiliated with Brazil's Roman Catholic Church, about 25,000 workers in Brazil are living in slavery-like conditions, most of them in the Amazon, and many of them working in the sugarcane business. The 2007 raid was not the first. In 2005, 1,000 workers were found living in debt slavery on a sugarcane plantation in the Brazilian state of Mato Grosso.[109]

Whether it is produced by slavery or not, the U.S. gobbles up a significant share of Brazil's ethanol output. In 2006, the U.S. imported about 400 million gallons of ethanol from Brazil.[110] That's equivalent to about 26,000 barrels per day. But remember, ethanol contains less heat energy than gasoline. Therefore, America's imports of Brazilian ethanol totaled only about 17,200 barrels per day of oil equivalent. That's not much when stacked next to America's 21-million-barrel-per-day oil habit.

Nor does that Brazilian ethanol matter much when compared with America's imports of Brazilian oil and oil products. In 2006, the U.S. imported an average of 133,000 barrels of crude per day from Brazil.[111] In addition to crude, Brazil also provides a variety of oil products to the U.S., including petroleum coke and fuel oil.[112] When all crude and petroleum products are accounted for, in 2006, the U.S. imported an average of 192,000 barrels of crude and oil products per day from Brazil.[113] In other words, in 2006, the U.S. imported 11 times as much energy from Brazil in the form of crude and oil products as it did in the form of ethanol.

FIGURE 10 Brazil's Oil Consumption, Oil Production, and Ethanol Consumption, 1965–2005

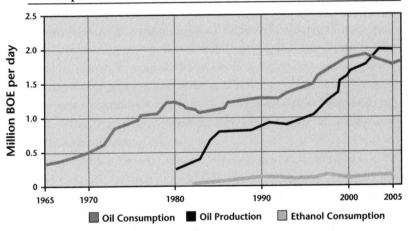

This graphic shows that Brazil's ethanol consumption has always been relatively small compared to its oil consumption and oil production.

Note: BOE stands for barrels of oil equivalent.

SOURCES: BP, Earth Policy Institute, EIA, Ethanol Producers Association.

Graphic courtesy of Tad Patzek, University of California, Berkeley.

The truth about Brazil's energy "miracle" is that it has almost nothing to do with ethanol and everything to do with Petrobras's ability to continue increasing its oil production—the vast majority of which is coming from Brazil's offshore waters. Since 1980, Petrobras has been growing its oil production by an average of 9 percent per year.[114] In 2006, that stunning success in the oil business allowed Brazil to become self-sufficient in terms of its oil needs.[115] And it's likely that Brazil's national oil company will continue ramping up its production. Petrobras hasn't fully exploited the Campos Basin, an offshore region east-northeast of the city of Rio de Janeiro. The Campos Basin holds an estimated 8.5 billion barrels of oil, or about 85 percent of Brazil's oil reserves.[116] To put the Campos Basin in perspective, it's larger than the giant East Texas field, which was discovered in 1930. Since its discovery, more than 30,000 wells have been drilled in the East Texas field.[117] So far, only about 1,000 oil and gas wells have been drilled in the Campos Basin.

In addition to Campos, Brazil has plenty of other offshore regions that await exploitation by Petrobras, which has become an elite player in the high-stakes casino of deepwater—and ultradeepwater—oil exploration. Between 1997 and early 2007, oil production at Petrobras more than doubled. By March 2007, the company was producing about 1.9 million barrels per day, most of which came from the Campos Basin.[118] By 2010, the company will likely be exporting 500,000 barrels of oil per day.[119] And by 2015, the company hopes to double its oil production again, to some 4.5 million barrels per day.[120] These plans gained credence in November 2007 when the company announced that its new offshore Tupi field may hold up to 8 billion barrels of oil equivalent. Tupi is the second-largest oil discovery in the last 20 years.

Within a few years, Brazil will likely be the biggest oil producer in Latin America, surpassing Venezuela, a traditional oil powerhouse (and OPEC member), which has seen its oil output plummet and its economy devastated under the rule of its populist president, Hugo Chávez.

Petrobras's amazing success is reflected in its stock price, which increased about 10-fold between late 2002 and early 2007.

FIGURE 11 Petrobras Stock Price, 2001–2007

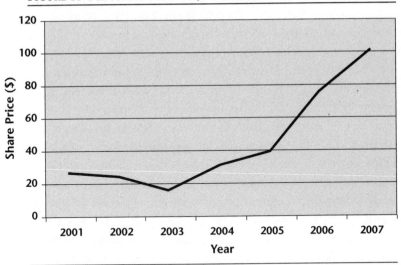

SOURCE: Price data from Marketwatch.com.
Note: Price data are not split adjusted. Chart shows only the price on the first trading day of each year.

Brazil's "miracle" has other factors as well, chief among them that Brazilian motor fuel is expensive. And those high prices have helped reduce consumption. Over the past decade and a half, Brazilian motorists have consistently paid higher taxes on their gasoline than their American counterparts, and those higher taxes have meant higher prices at the pump. In 1993, Brazilians paid $0.53 per liter of gasoline, while Americans paid $0.32. In 1998, they paid $0.80, while Americans paid $0.32. In 2004, they paid $0.84 per liter, while Americans were paying just $0.54.[121] In early May 2007, Brazilians were paying up to $1.38 per liter (about $5.22 per gallon), while Americans were paying about $0.82 per liter (about $3.09 per gallon).[122]

These higher prices, along with Brazil's smaller population and smaller fleet of vehicles, help explain why Brazil's oil consumption is relatively small: about 2.1 million barrels of oil per day, or about one-tenth as much as the amount used by the U.S.[123] On a per capita basis, the average Brazilian uses about 0.5 gallons of oil per day.[124] The average American uses six times as much—about 3 gallons per day.

Khosla and Daschle could replicate Brazil's "miracle" in the U.S. if they could just convince every American to use as little oil as the average Brazilian does. If that were to occur, the U.S. would immediately be energy independent or, at least, oil independent. Here's why: America uses 21 million barrels of oil per day. One-sixth (16 percent) of that amount would be 3.3 million barrels per day. The U.S. produces 5.1 million barrels of crude per day.[125] Presto! America would suddenly be oil independent. And all it took to achieve oil independence was a federal mandate requiring an 84 percent reduction in oil consumption for every American. In approximate terms, that would mean that Americans would be allowed to drive their cars or fly in airplanes only one day per week. Along with oil independence, the U.S. could quickly become one of the biggest oil exporters in the Western Hemisphere, on a par with Venezuela.

Brazil's ethanol sector is an important part of that country's economy, but it's not nearly as important as oil. That can be proven by following the money. In 2007, Petrobras announced that it wanted to increase its ethanol production and export capacity. Toward that end, by 2012, it will spend about $1.6 billion upgrading its ethanol production, storage, and transportation capabilities.[126] That's a substantial investment. And yet, during that same time period, Petrobras will spend $38 billion—about 24 times as much—on its oil and gas exploration and production activities.[127]

Although politicos and pundits continue to obsess about Brazil's ethanol business, the true picture is far different. Brazil will not need any imported oil, not because of ethanol, but because of the vast offshore oil reserves it has begun to tap. That oil has made Brazil into one of the most important energy producers in the Western Hemisphere.

Despite the truth about Brazil's oil, politicians and others continue focusing on Brazil's ethanol. And the myth of Brazil's ethanol miracle bears a striking resemblance to another oft-repeated fable: Cellulosic ethanol will soon be displacing huge amounts of gasoline.

CELLULOSIC ETHANOL:
THE VAPORWARE OF THE ENERGY SECTOR

If cellulosic ethanol were part of the computer industry, it would fall under the category of "vaporware." That is, it's an idea that gets promoted and promised over and over, but it's never actually delivered.

No other fuel in modern American history has gotten more hype while providing fewer actual Btus than cellulosic ethanol. (Hydrogen probably comes in a close second.)

The idea is that farmers will grow "energy crops," which will then be fed into a distillery that employs enzymes or some other yet-to-be-developed process to turn that crop into ethanol. The promise of cellulosic ethanol is a favorite topic for people like Amory Lovins and James Woolsey, both of whom promote it at every turn. In his 2004 book *Winning the Oil Endgame*, Lovins declared that advances in biotechnology will make cellulosic ethanol viable and that replacing hydrocarbons with carbohydrates "will strengthen rural America, boost net farm income by tens of billions of dollars a year, and create more than 750,000 new jobs."[128] In 2006, before the U.S. Senate, Lovins testified that the U.S. could dramatically cut its oil consumption by using more natural gas "and advanced biofuels (chiefly cellulosic ethanol) for the remaining oil at an average cost of $18 per barrel."[129] (By any measure, Lovins's estimate is absurdly low. Producing ethanol for $18 per barrel implies production costs of just $0.43 per gallon. That's about one-fourth as much as the cost of producing gasoline in mid-2007 at a major oil refinery on the Houston Ship Channel.)

Woolsey has been talking about cellulosic ethanol since at least 1999, when he and U.S. Senator Richard Lugar, an Indiana Republican, wrote that cellulosic ethanol would "democratize the world's fuel market."[130]

A panoply of others have bought into the cellulosic ethanol hype. In 2004, the neocons' Set America Free plan said that America should increase ethanol output "by commercializing the production of ethanol from biomass waste and dedicated energy crops."[131]

After his 2006 movie, *An Inconvenient Truth,* was released, former vice president Al Gore promised that cellulosic ethanol would be "a huge new source of energy, particularly for the transportation sector. You're going to see it all over the place. You're going to see a lot more flex-fuel vehicles. You're going to see new processes that utilize waste as the source of energy, so there's no petroleum consumed in the process."[132] Gore's former boss, Bill Clinton, loves cellulosic ethanol, too. While promoting Proposition 87 in California in 2006, he declared, "These things are not expensive. We have this kind of biomass to make cellulosic ethanol all over America. It would increase income in rural America. It would increase income in rural California. It would stabilize the environment and improve our national security."[133]

John Kerry, another former Democratic presidential aspirant, claims that cellulosic ethanol "has the potential to substantially reduce our consumption of gasoline. . . . A technological breakthrough could lead to widespread use of cellulosic ethanol to fuel our vehicles."[134]

In his 2006 State of the Union speech, George W. Bush said the federal government will "fund additional research in cutting-edge methods of producing ethanol, not just from corn, but from wood chips and stalks, or switch grass. Our goal is to make this new kind of ethanol practical and competitive within six years."[135] A year later, in his 2007 State of the Union speech, Bush once again touted the fuel, saying that the U.S. "must continue investing in new methods of producing ethanol—using everything from wood chips to grasses, to agricultural wastes."[136]

In February 2007, Bush visited an enzyme production facility in North Carolina owned by a company called Novozymes. In a speech at the plant he said, "It's an interesting time, isn't it, when you're able to say, we're on the verge of some breakthroughs that will enable a pile of wood chips to become the raw materials for fuels that will run your car."[137] That same day, the White House put out a "fact sheet" declaring that as "cellulosic ethanol production becomes commercially viable," plants around the country will be able to use "grass from a prairie, wood chips from a forest, or agricultural waste like stalks—to create fuel."[138]

In May 2007, shortly after the Intergovernmental Panel on Climate Change released another of its reports, the editors of the *New York Times* ran an opinion piece saying that "we can and must" begin "investing in alternative fuels, like cellulosic ethanol, that show near-term promise."[139] In September 2007, the *Times* ran another editorial saying that cellulosic ethanol would "confer major benefits in terms of oil savings and reduced greenhouse gases."[140]

Cellulosic ethanol clearly has a church-full of believers. But belief doesn't put fuel in your tank.

Human beings have been turning food into alcohol for about 6,000 years.[141] And they have gotten remarkably good at it. Humans have learned to make alcohol from apples, grapes, corn, rice, wheat, honey, potatoes, barley, sugarcane, and lots of other food products. But over those 6,000 years of experimentation, fermentation, and drunkenness, they've seldom had much success at turning sawdust, grass, or wood chips into alcohol. The reason is simple: Plant cellulose is a lousy feedstock for making alcohol.[142]

Cellulose is difficult to convert into alcohol because it is actually a polymer made up of chains of sugar molecules. In that regard, it is similar to the starch found in corn. But unlike starch, the sugar chains in cellulose are packed together like thin sticks in a bundle. That bundled structure gives plant walls the stiffness they need to form strong stalks and branches.

The challenge in making cellulosic ethanol comes in trying to free the sugars that are bound up in the plant walls. As *Chicago Tribune* reporter Michael Oneal wrote in an excellent October 13, 2006, article on cellulosic ethanol, "Breaking down those walls is like robbing a bank. While the starch in corn kernels gives up its energy-packed sugars easily, the sugars in plant cell walls are locked into winding structures of complex carbohydrates designed to give plants backbone and protection."[143]

Robbing the cellulose bank of its sugary riches requires that steam or acid be used. Enzymes may also be added to the cellulose and water mixture to assist in this process. After the cellulose/water mixture is fermented, the alcohol must be stripped out (distilled) from the mixture. But there's a key difference between the process used to make grain alcohol and the one used to produce cellulosic ethanol. Grain alcohol fermentation produces a grain/water mixture (distillers call it *mash*) that contains from 15 to 20 percent ethanol. Heat is then used to distill the ethanol out of the mash. In the cellulosic fermentation process, the ethanol content in the mash is only about 5 percent. That means that stripping the ethanol from that larger volume of mash requires far more energy than the distillation needed to produce ethanol from grain.[144]

Another key problem: the Bunyanesque volumes of material needed to make cellulosic ethanol. In Oneal's *Chicago Tribune* article, one entrepreneur working on plant genetics described this problem as the "tyranny of distance." That is, the further a company has to haul the huge bundles of switchgrass or corn stubble or straw, the more expense is incurred.

According to the U.S. Department of Energy, a ton of biomass can produce about 80 gallons of ethanol.[145] Let's suppose the objective is to build a plant that can produce 80 million gallons of ethanol per year from corn stubble. That amount would require 1 million tons of stubble. According to calculations done by Bill Hord, a reporter at the *Omaha World-Herald* newspaper, hauling that amount of material would require "67,000 semitrailer loads to haul the baled stubble out of the field. That's 187 truckloads a day, or one every eight minutes. To complicate matters, the need for trucks, machinery and manpower would come during harvest, already the busiest time of the year on the farm." Hord continued, saying that an ethanol distillery that used 2,000 tons of corn stubble (also called stover) per day would require "100 acres stacked 25 feet high with stover to run a refinery for a year."[146]

And what would be the result of all that effort? Those 80 million gallons of hard-won cellulosic ethanol would be the equivalent of about

53 million gallons of gasoline, or about 0.04 percent of America's annual gasoline consumption.[147]

Hord's calculations were seized upon by Robert Rapier, the writer of the R-Squared Energy Blog.[148] Rapier mockingly wrote that if all the biomass inputs needed to make cellulosic ethanol from corn stubble were free, "all we would need is 2,500 of these [80-million-gallon-per-year] facilities, and we will have met all U.S. gasoline needs (but not diesel, fuel oil, or jet fuel). Ah, but we forgot about energy inputs. How many gallons of fossil fuels did it take to run all of those semi-trailer trucks to take the stubble to the plant? How much natural gas was required to distill off the ethanol?"[149]

Rapier's point goes to the heart of the matter: How much energy profit can be made with cellulosic ethanol? Patzek and Pimentel estimate that cellulosic ethanol production is an energy loser, yielding net energy losses of about 50 percent.[150] That is, for 1 Btu invested, an investor gets 0.5 Btus in return.

Obviously, the gargantuan volumes of biomass needed to produce cellulosic ethanol will be hard to manage. And that fact has led companies to work on developing new genetic strains of plants with high energy content so that the hauling and storage costs are compensated by the calorific value of the plants.[151] But try as they might, despite decades of research, there haven't been enough breakthroughs to overcome all of the hurdles.

The result is abundantly clear: Cellulosic ethanol remains far too expensive to produce on a large scale. That fact has been established by numerous studies. For instance, an August 2006 study by a group of California-based scientists who specialize in carbon sequestration estimated that the cost associated with turning waste corn stalks into cellulosic ethanol was "twice as high as that of ethanol from corn. Forest residues and wastes, biomass crops, and municipal wastes are even less promising. The conclusions of this assessment are that none of the existing processes are ready for commercial applications in any foreseeable time frame and that continuing fundamental and applied R&D is required."[152]

In late 2006, the administrator of the Energy Information Administration, Guy Caruso, said that the capital costs associated with producing cellulosic ethanol were five times as much as those associated with corn ethanol production.[153] In January 2007, researchers at Oregon State University also concluded that cellulosic ethanol was too expensive. They estimated that the cost of ethanol produced from wood was "nearly 200 percent higher than gasoline."[154] In May 2007, researchers at Iowa State University's Center for Agricultural and Rural Development found that cellulosic ethanol is not "economically viable in the Corn Belt under any of the scenarios" that they looked at. And like Hord, they determined the cost of transporting the cellulose to be prohibitive: "Cellulosic ethanol from corn stover does not enter into any scenario because of the high cost of collecting and transporting corn stover over the large distances required to supply a commercial-sized ethanol facility."[155]

There are plenty of other doubters. In May 2006, in an opinion piece published by the *Wall Street Journal,* former CIA director John Deutch, who's now a chemistry professor at the Massachusetts Institute of Technology, wrote that biofuels should be considered along with other technologies as a replacement for oil-based fuels. But he was cautious about the prospects for cellulosic ethanol. He concluded that producing enough ethanol from switchgrass to displace 1 million barrels of oil per day would require that 25 million acres of land—about 39,000 square miles—be planted in switchgrass. That's an area the size of Kentucky.[156] While Deutch encouraged scientists to continue looking for a breakthrough in the development of cellulosic ethanol, he wrote, "I conclude that we can produce ethanol from cellulosic biomass sufficient to displace one to two million barrels of oil per day in the next couple of decades, but not much more. This is a significant contribution, but not a long-term solution to our oil problem." [157]

If Deutch's estimates are correct, producing enough cellulosic ethanol to displace just *half* of America's daily ration of 21 million barrels of oil per day would require the U.S. to plant an area approximately equal to 1.5 times the size of Texas.[158]

Despite all the blather about cellulosic ethanol, the technology remains insufficient to make it viable. Back in 1997, the Government Accountability Office concluded that "cellulose-based ethanol remains an experimental rather than an economically viable technology."[159]

That's still true today.

Nevertheless, the Department of Energy has handed out $385 million in grants to about a half dozen companies who are working on cellulosic ethanol.[160] And those companies might have some success. But the punch line here is obvious: Start believing the cellulosic ethanol evangelicals when you can actually pump cellulosic ethanol—at a competitive, unsubsidized price—into your fuel tank.

ETHANOL CRIMPS GASOLINE PRODUCTION

No one likes to feel sympathy for the big refiners. But the federal mandates on the use of ethanol, combined with the phaseout of another motor fuel additive, MTBE, has caused significant problems for the refining industry, and those problems are resulting in higher prices for consumers.

For years, MTBE (methyl tertiary-butyl ether) was used in the U.S. as a gasoline additive. (It is still used in numerous countries around the world.) MTBE is a fuel oxygenate. Oxygenates help gasoline burn more cleanly and therefore can improve air quality by reducing toxic emissions, particularly carbon monoxide. In the mid-1990s, the federal government began requiring refiners to use oxygenates in gasoline that was sold in cities with air quality problems. Refiners liked MTBE because its energy content was similar to that of gasoline and it was not overly volatile—that is, it didn't evaporate quickly. Those qualities led refiners to add up to 15 percent MTBE to the reformulated gasoline that they were supplying to areas of the U.S. that had air quality problems.

But due to some leaking underground storage tanks, MTBE-blended gasoline was also causing some groundwater problems. And those leaks led to lawsuits over the refiners' use of the additive. By

2003, 7 states had banned the use of MTBE, and by 2006, 10 other states had also banned the substance.[161] Congress was reluctant to provide any legal indemnity for the companies that used the additive. In mid-2006, the industry quit using MTBE and turned to ethanol as the oxygenate of choice.[162]

Like MTBE, ethanol is a good oxygenate. But it creates big problems for oil refiners because it increases the volatility of gasoline. Throughout the year, refiners must adjust the volatility of their fuel in order to compensate for the changes in air temperature. In summer, you want gasoline to evaporate slowly so that the fuel doesn't cause air quality problems. In winter, you want it to evaporate quickly so that it will readily ignite inside the car's engine.[163] Therefore, during the summer months, in order to compensate for the ethanol, refiners have to strip out the lighter hydrocarbon fractions from their gasoline (usually propane, butanes, and pentanes) in order to comply with federal air quality rules. Losing those constituents means they cannot produce as much gasoline from the same amount of crude oil. This loss of production varies, but it is about 5 percent of the total gasoline volume.[164] In 2005, the Government Accountability Office issued a report on the domestic gasoline market that addressed the loss of volume, saying, "Removing these components and reprocessing them or diverting them to other products increases the cost of making ethanol-blended gasoline."[165]

Ethanol-blended gasoline means refiners can make less gasoline out of each barrel of crude. Lower production volumes mean less marketable product. And less marketable product means higher prices for consumers at the pump.

ETHANOL COSTS MORE AT THE PUMP

In October 2006, *Consumer Reports* ran tests on a Chevrolet Tahoe SUV and found that its fuel economy dropped by 27 percent when the vehicle was running on E85 (a blend of 85 percent ethanol and 15 percent gasoline) compared to regular gasoline. The magazine calculated that with E85 selling for $2.91 per gallon, consumers would have to spend $3.99 on E85 in order to get the energy equivalent of one gallon of gasoline.[166] But even though ethanol is inferior to gasoline when it comes to Btu content, it's still more expensive to make. In February 2007, the Government Accountability Office issued a report declaring flatly that while corn ethanol production is "technically feasible, [corn ethanol] is more expensive to produce than gasoline."[167]

Given its higher production costs, it's not surprising that ethanol is more expensive. Between 1982 and 2006, the price of ethanol was *always* higher than the price of gasoline. That was true even though ethanol, given its lower heat content, should sell for about 66 percent of the cost of an equivalent amount of gasoline. And yet, it never has.

Despite this price history, Bob Dinneen, the president of the Renewable Fuels Association, the lobbying group for the ethanol industry,

FIGURE 12 Ethanol and Unleaded Gasoline Rack Prices FOB Omaha, Nebraska, 1982–2006

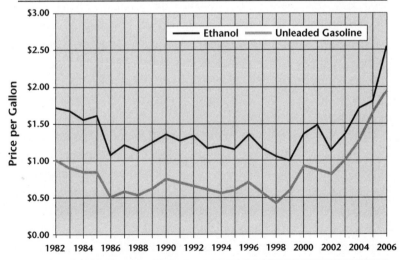

Source: Official Nebraska Government Web site, "Ethanol and Unleaded Gasoline Average Rack Prices." Available: http://www.neo.ne.gov/statshtml/66.html. Note: FOB stands for free on board, which is the price at the loading point.

claimed in a letter to the *Wall Street Journal* in May 2007 that "consumers will benefit significantly as biofuels drive down the price of gasoline."[168] Dinneen made that spurious claim even though after more than two decades of subsidies to ethanol producers, consumers have yet to see *any* benefit at the pump from biofuels.

Not only are consumers paying more at the pump thanks to congressional mandates requiring ethanol production, but they are also breathing dirtier air.

ETHANOL HURTS AIR QUALITY

The Environmental Protection Agency's Web site says its mission is "to protect human health and the environment."[169] And yet, when it comes to ethanol, the agency has stated in very clear language that increased use of ethanol in gasoline will mean *worse* air quality in America.

Of course, that's not the official story.

In an April 10, 2007, press release announcing the Renewable Fuel Standard—the federal program mandated by Congress when it passed the Energy Policy Act of 2005 that requires U.S. refiners to be blending at least 7.5 billion gallons of ethanol per year into America's motor fuel supply by 2012—EPA administrator Stephen L. Johnson declared that the use of more ethanol "offers the American people a hat trick—it protects the environment, strengthens our energy security, and supports America's farmers."[170]

The only problem is that on that same day, Johnson's agency issued a fact sheet saying that using more ethanol will result in major increases in the release of two of the worst air pollutants: volatile organic compounds and nitrogen oxides. The fact sheet on the Renewable Fuel Standard says, "Nationwide, EPA estimates an increase in total emissions of volatile organic compounds and nitrogen oxides ($VOC + NO_x$) [of] between 41,000 and 83,000 tons." It goes on, saying, "Areas that experience a substantial increase in ethanol may see an increase in VOC emissions between 4 and 5 percent and an increase in NO_x emissions between 6 and 7 percent from gasoline powered vehicles and equipment."[171]

When asked about the higher emissions, an EPA spokesperson, Jennifer Wood, insisted that the increased NO_x and VOC numbers were "very minimal increases." She also told me that the agency has other "tools under the Clean Air Act to reduce NO_x."[172]

Wood's claim that the ethanol-related increases in pollution are "minimal" leaves clean air advocates like William Becker of the National Association of Clean Air Agencies gasping. He said the EPA is "scoffing at a 4 to 7 percent increase in air emissions at a time when agencies across the country would do anything to achieve that kind of a reduction in VOCs and NO_x." Becker's group represents air pollution control authorities from 49 of the 50 states and several territories, as well as local agencies from 165 metro areas around the U.S. He said the 4 to 7 percent increases admitted by the EPA are "a significant amount of emissions in any location in this country. And we can't just willy nilly be giving it away, particularly when states are struggling to meet current ozone standards."[173]

The negative health effects of ethanol-blended gasoline have placed the EPA in the odd position of enforcing rules that run directly counter to its stated goals. On its Web site, the agency says, "Reducing emissions of NO_x is a crucial component of EPA's strategy for cleaner air."[174] And the agency's Web site makes it clear why it wants to reduce NO_x emissions: NO_x can cause ground-level ozone, acid rain, increases in particulate matter, water pollution, the unleashing of toxic chemicals, a reduction in visibility, and climate change.[175] The agency also explains in very clear terms that VOCs lead to the creation of ground-level ozone, one of the most dangerous urban pollutants. Ozone is created when NO_x and VOCs are mixed in the presence of sunlight.

The EPA's Web site gives this primer on ozone:

Breathing ozone can trigger a variety of health problems including chest pain, coughing, throat irritation, and congestion. It can worsen bronchitis, emphysema, and asthma. Ground-level ozone also can reduce lung function and inflame the linings of the lungs. Repeated exposure may permanently scar lung tissue. Ground-level ozone also damages vegetation and ecosystems. In the United States alone, ozone is responsible for an estimated $500 million in reduced crop production each year.[176]

Other regulators and scientists, not just the EPA, are also finding that ethanol worsens air quality.

California regulators and politicians have been jousting with the EPA over ethanol for nearly a decade due to concerns about air quality. Back in March 1999, the state of California banned the use of MTBE because of worries about possible groundwater contamination. About that same time, due to concerns that ethanol would worsen air quality, the state's governor, Gray Davis, formally asked the EPA for a waiver from the federal mandate requiring that ethanol be added to the state's gasoline. But in 2001, the EPA denied the state's request.[177] Ever since then, the state has been battling the agency over the ethanol requirement.

In 2004, the California Air Resources Board released a study that found that gasoline containing ethanol caused VOC emissions to increase by 45 percent when compared to gasoline containing no oxygenates.[178] In mid-2006, the South Coast Air Quality Management District, the agency that handles air quality issues for some 15 million people living in or near Los Angeles County, determined that gasoline containing 5.7 percent ethanol may add as much as 70 tons of VOCs per day to the state's air.[179] And that's on a day when the temperature is just 83 degrees. When the thermometer gets closer to 100 degrees, the evaporative emissions could be as much as 118 tons of VOCs per day. In a meeting held to discuss these findings, one official from the air district agency noted that a pollutant increase of that size was gargantuan, particularly given that an air quality regulator in the region around Los Angeles "can become employee of the month by coming up with a way of reducing emissions by one-tenth of a ton per day."[180]

In April 2007, Mark Z. Jacobson, an associate engineering professor at Stanford University, published a study finding that the widespread use of E85 "may increase ozone-related mortality, hospitalization, and asthma by about 9 percent in Los Angeles and 4 percent in the United States as a whole" when compared to the use of regular gasoline. The study also found that due to its ozone-related effects, E85 "may be a greater overall public health risk than gasoline."[181]

University of Denver chemistry professor Donald Stedman has been studying ethanol's effects on air quality for about 15 years. He believes that the EPA's estimates of the potential increases in VOCs and NO_x are too low. Further, Stedman says that any increase in VOCs will hurt air quality. "You increase VOCs, you increase ozone. Period. For everybody, everywhere," he told me in April 2007. "It's a disaster. And I think it [the increased use of ethanol in gasoline] will raise VOC emission more than the 5 to 6 percent that EPA is admitting."[182]

Studies that Stedman did in Denver showed a correlation between ethanol and worsening air quality. Stedman found that as ethanol gained market share in the Denver area, ozone levels rose. His explanation: The increased vapor pressure in ethanol-blended gasoline allowed more VOCs to get into the air, and those VOCs turned into ozone.

The South Coast Air Quality Management District's findings appear to support Stedman's claim that the real-world emissions of VOCs will exceed the EPA's projections. If the air district agency is correct, and VOC emissions increase by 70 tons per day, then California alone would see an increase of 25,550 tons of VOCs per year due to ethanol-blended gasoline. That's more than half of the 41,000 ton-per-year nationwide minimum that the EPA projected in its April 10, 2007, fact sheet.

So exactly why is the EPA advocating the increased use of ethanol when its own studies have shown that ethanol use will worsen air quality for millions of Americans? Well, a possible explanation can be found in the final line of the agency's April 11, 2007, press release regarding the rules on the Renewable Fuel Standard. The line explains that, thanks to the increased use of ethanol, "net U.S. farm income is estimated to increase by between $2.6 and $5.4 billion."[183] Precisely why the EPA, an agency charged with protecting the environment, cares about the level of U.S. farm income was not discussed.

Alas, the EPA hasn't bothered to analyze how more ethanol production will affect America's water supply. If it did, the agency's regulators might be shocked.

ETHANOL PRODUCTION REQUIRES
ENORMOUS QUANTITIES OF WATER

Just four months before the EPA released the fact sheet showing that more ethanol use will make air quality worse, scientists at Sandia National Laboratory in New Mexico issued an 80-page report called "Energy Demands on Water Resources."

The December 2006 report looked at all aspects of water use and its relationship to energy production. For instance, it explains that the electric power industry is a huge water consumer, using about 39 percent of all the freshwater in the U.S., or about as much as all of the irrigated farmland in the country.[184] But the real shock in the Sandia report comes from its discussion of corn ethanol. In a section on biofuels, the report says that the amount of water needed to grow corn varies widely, but that the amount of "water required for production of irrigated corn is 11,000 gal per MMBtu."[185] In layman's terms, that means that ethanol produced from irrigated corn requires 11,000 gallons per million Btus of fuel produced, or about 880 gallons of water for every gallon of ethanol. Add in another 5 gallons of water needed at the distillery to turn the corn into ethanol, and each gallon of ethanol requires 885 gallons of freshwater.[186] For comparison, the average bathtub holds 35 gallons.[187] Thus, a gallon of ethanol produced from irrigated corn requires as much water as the amount contained in 25 bathtubs.

Of course, not all corn is irrigated. By some estimates, just 15 percent of all U.S. corn is produced with irrigation.[188] But even if that is the case, local demands for water to produce corn ethanol can be extraordinary. For instance, some 70 percent of the corn grown in Nebraska—the third-largest corn-producing state—relies on irrigation.[189]

Furthermore, even if just 15 percent of the corn in the U.S. is produced with irrigation, ethanol production is still hugely water-intensive. The math is clear: Assume that 15 percent of the corn used during the production of America's ethanol in 2006 came from irrigated fields. In that case, each gallon of domestic ethanol required the consumption of about 132 gallons of water.[190] That's a huge quantity, particularly when compared to the quantity needed for oil and gas production. According to the Sandia report, the extraction and refining of conventional oil requires—at most—2.8 gallons of water for each gallon of oil produced.[191] (The report puts the minimum water consumption at about 1.3 gallons per gallon of refined product.)

The Sandia scientists found that at nearly every step of the production cycle, ethanol requires more water than oil-based fuels. They determined that the distilling process requires at least two times as much water as the oil-refining process. The report says that distilling corn into ethanol takes nearly 5 gallons of water for every gallon of ethanol produced.[192] Thus, ethanol requires between two and four times as much water during the refining process as oil.

The results of the Sandia report were corroborated by a June 2007 study (cited earlier) by Jan Kreider and Peter Curtiss. In fact, they found that the Sandia estimates of ethanol's water needs were too low. Kreider and Curtiss claim that each gallon of corn ethanol requires 170 gallons of water. For comparison, the two scientists estimated that each gallon of gasoline requires just 5 gallons of water. If Kreider and Curtiss are right, the 5 billion gallons of corn ethanol that were produced in America in 2006 used more water than all of the water that was used during the production of all of the gasoline consumed in the U.S. that year.

In Kreider and Curtiss's estimate of 170 gallons of water per gallon of ethanol, those 5 billion gallons of ethanol required the use of 850 billion gallons of water. By contrast, the production of 140 billion gallons of gasoline (America's consumption in 2005) required the use of 700 billion gallons of water. So how much is 850 billion gallons of water? Well, it's equal to about 2.6 million acre-feet, or about as much water as is contained in Lake Texoma, the largest lake, by volume, in Oklahoma.[193]

Ethanol's water demands will grow right along with its increasing production. For instance, if the U.S. managed to reach George W. Bush's goal of 35 billion gallons of alternative-fuel production per year, and all of that fuel was corn ethanol, that amount would require the consumption of 18.2 million acre-feet of water per year. That's nearly half as much water as the 36 million residents of California use per year.[194]

Kreider and Curtiss also looked at the water needs for other fuels. Perhaps most intriguing is the calculation that they did for cellulosic ethanol. They determined that cellulosic ethanol's water needs are nearly as great as those for corn ethanol, with 146 gallons of water needed for each gallon of cellulosic ethanol produced. If their assumptions are correct, cellulosic ethanol's water demand is about 29 times greater than that of conventional gasoline.

These calculations are important because cellulosic ethanol has repeatedly been sold as one of the panaceas for America's energy needs. So let's assume that corn ethanol fades away and is replaced by cellulosic ethanol, and that it is used to meet all of Bush's target of 35 billion gallons of renewable and alternative fuels produced in the U.S. by 2017. In that scenario, the water needed to provide that much cellulosic ethanol would equal some 5.1 trillion gallons of water per year, or about 15.7 million acre-feet. That's approximately equal to the entire annual flow of the Colorado River, one of the main sources of water for cities in the southwestern U.S. and the seventh-longest river in America.[195]

Kreider and Curtiss report that soybean-based biodiesel is even more water-intensive, requiring 900 gallons of water for each gallon of biodiesel produced.[196] (For a table from Kreider and Curtiss's report that compares various fuels, their land use, their water demand, and other factors, see Appendix C.)

The extraordinary amounts of water needed for biofuel production will exacerbate water shortages in the Plains and western states, which are already experiencing excess water demand. The Ogallala aquifer, which underlies parts of Colorado, Kansas, Nebraska, New Mexico, Oklahoma, and Texas, has been heavily used for agricultural irrigation and has already been depleted by about one-half. Recharge from rainfall is supplying only a fraction of the amount of water that is being withdrawn from the aquifer.[197]

FIGURE 13 Water Demand for Production of Various Motor Fuels

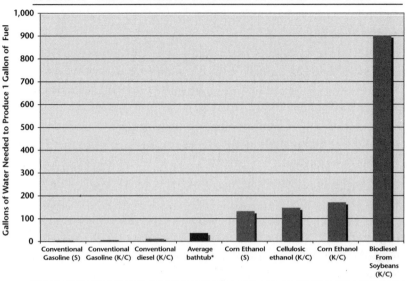

Note: *Bathtub shown for scale purposes only. S: Sandia National Laboratory report, 2006. USDOE, op cit. K/C: Jan F. Kreider and Peter S. Curtiss, "Comprehensive Evaluation of Impacts from Potential, Future Automotive Fuel Replacements," *Proceedings of Energy Sustainability 2007*, June 27–30, 2007.

Residents in several small towns around the country are opposing ethanol distilleries because of their enormous thirst for water. In 2007, opponents of a proposed $165-million ethanol plant in Rogersville, Missouri, pointed out that the plant would require some 1.3 million gallons of groundwater per day.[198] In Wright, Kansas, area residents opposed an ethanol plant that was expected to use 1.5 million gallons of water per day. But it wasn't the plant itself that worried the opponents, it was the amount of water needed to grow the corn. The Associated Press quoted one local businessman who said that "what's really going to kill the water situation here" will be "the amount of water it will take to irrigate all that corn."[199]

In April 2007, the Kansas Water Authority declared that the decline of the Ogallala needs to be a "higher priority" for the state's water officials.

One of the key factors cited by the authority's board members was their concern about ethanol and its potential to soak up critical water resources in western Kansas.[200]

Many other examples could be cited here of the problems associated with water supply and ethanol. It's abundantly clear that the U.S. simply doesn't have enough water to sustain massive increases in ethanol production. One of the most surprising conclusions in the Sandia report can be found on its concluding pages. The Sandia scientists said that higher levels of fuel imports might be good for the U.S., when it comes to water availability. They wrote, "An increase in imports, especially of refined fuels, could lead to a decrease in domestic water needs for fuels."[201]

That makes perfect sense. And it underscores the fundamental idiocy of the entire ethanol and energy independence effort. Compared to the U.S., the oil-rich countries of the Persian Gulf have very little freshwater and, therefore, little ability to grow food. By comparison, the U.S. has relatively large amounts of water and a tremendous ability to grow food. This means that in the global market, America's ability to import oil from the Persian Gulf may actually be an advantage. "We exchange our food for their fuel and both parties benefit," says Tad Patzek, the University of California-Berkeley engineering professor who is among the world's best-known ethanol critics. But if the U.S. continues trying to turn its food into fuel, he warns, "we will bankrupt ourselves with regard to water."

FLEX-FUEL VEHICLES AND THE E85 SCAM

The great comic actress Lily Tomlin once said that "no matter how cynical I get, I can't keep up." After watching George W. Bush's March 2007 appearance at the White House with the top executives from America's biggest automakers, it's easy to identify with Tomlin's comment.

During the meeting, the automakers were given an opportunity to display their latest vehicles on the White House driveway. The cars were all flex-fuel vehicles (FFVs) that were capable of burning E85. The media event was held to demonstrate the automakers' and the Bush administration's commitment to reducing oil consumption. During a short speech in front of the shiny new cars, Bush declared that it was important to reduce gasoline consumption "for national security reasons." Promoting cars that burn ethanol, he said, was the way to accomplish that goal. After Bush spoke, the CEO of General Motors, Rick Wagoner, said that his company could see a "path" toward "lower amounts of imported oil." And that path, he said, was going to be driven by flex-fuel vehicles that can use E85. Wagoner was followed by Alan Mulally, the CEO of Ford Motor Co., who said that Ford was supportive of Bush's goal "both for energy efficiency and independence." He was followed by Tom LaSorda, the CEO of DaimlerChrysler, who said that his company believes that ethanol "is the answer for America to lower our dependence on foreign oil."[202]

After giving their statements, neither Bush nor the auto bosses took any questions from the assembled media.

A hard look at the FFV program shows why questions were not welcomed. It also demonstrates why Lily Tomlin just can't get cynical fast enough. Although the CEOs declared their desire to cut oil use, the automakers' embrace of ethanol has produced a result that is the exact opposite of their stated goal. That is, by building more ethanol-capable cars, the big Detroit automakers have actually *increased* oil consumption in America.

Here's how.

Over the past decade or so, Ford Motor Co. has built more than 1.6 million vehicles capable of burning gasoline as well as motor fuel containing up to 85 percent ethanol (E85).[203] Over that same period, General Motors has built more than 1.5 million of these flexible-fuel vehicles.[204] DaimlerChrysler has sold about 1.5 million them.[205]

To woo consumers, General Motors has created an entire marketing program around its FFVs. In 2006, during the Olympics, the giant automaker launched its "live green, go yellow" campaign to promote its ethanol-capable FFVs. In announcing the promotion, a GM spokesman said that ethanol is "a great energy option" and that "in the world

President George W. Bush talks to the media after a demonstration Monday, March 26, 2007, of alternative-fuel vehicles on the South Lawn drive of the White House. Standing with him, from left, are Rick Wagoner, chairman and CEO, General Motors Corporation; Alan Mulally, president and CEO, Ford Motor Co.; Tom LaSorda, President and CEO, DaimlerChrysler Corporation; and Secretary of Transportation Mary Peters.

PHOTO CREDIT: White House photo by Joyce Boghosian. Photo available: http://www .whitehouse.gov/news/releases/2007/03/images/20070326_p032607jb-0065 -515h.html.

of ethanol, yellow is the new green, since today its main source is from corn."[206] DaimlerChrysler has declared that its FFVs are part of the "political target of reducing the American dependence on foreign oil."[207] Ford tells potential buyers that the "best part about driving an FFV is that it doesn't require a lifestyle change. Just fuel up with E85 or regular gasoline—whichever is more convenient."[208]

But E85 is not convenient at all. In fact, it's a scarce commodity. By early 2007, when Bush was chatting up the automakers at the White House, only about 900 of the 170,000 service stations in America were selling E85. In the entire state of California, there were only three service stations selling E85. In New Jersey, there were none at all.[209] Some ethanol boosters insist that the solution to using less foreign oil is to make E85 available at more service stations. In 2006, former presidential candidate John Kerry gave a speech in Boston in which he said that the U.S. should "immediately expand our investment in E85 infrastructure. Mandate that 10 percent of all major oil company filling stations offer at least one ethanol pump by 2010."[210] But that won't be cheap. Adding an E85 pump costs up to $200,000 per service station.[211] Even if there were more E85 available for purchase, it wouldn't mean that consumers would be able to use it. In 2008, less than 3 percent of the motor vehicles in America were capable of burning E85.[212]

So why are Detroit's biggest automakers producing so many vehicles that can burn E85 when the vast majority of motorists in America cannot even buy it?

The answer reveals one of the most outrageous aspects of the ethanol scam: By producing FFVs, the automakers get credits from the federal government on the overall efficiency of their fleets. And therein lies the essence of the FFV scam: When calculating fuel efficiency for a given vehicle, the federal government counts only the amount of *gasoline* that it consumes.

The automakers claim that an FFV will be burning E85, not gasoline, for part of the time. That allows them to inflate the fuel efficiency numbers that they must meet under the federal government's Corporate Average Fuel Economy standards. For instance, GM seized on the federal credits by making one of its biggest SUVs, the Suburban, into an FFV capable of using E85. In the real world, the Suburban gets less than 15 miles per gallon. But thanks to the credits, the E85-capable Suburban is magically transformed into a vehicle that gets more than 29 miles per gallon.[213] Of course, that mileage occurs only on paper, not on the highway.

Thanks to the E85-FFV scam, the Sierra Club estimates that the big Detroit automakers raised the overall fuel efficiency of their fleets—again, *on paper*—by as much as 1.2 miles per gallon.[214] And they continue churning out huge numbers of FFVs, even though 99 percent of the E85-capable FFVs that they build will never burn E85. And they do so because it's cheap: The parts needed for an E85-capable FFV add only about $100 to the overall production costs for each vehicle.[215]

Building lots of E85-capable FFVs has allowed both Ford and GM to avoid fines from the federal government. According to an April 2, 2007, story in the *Wall Street Journal*, in three consecutive years—2003, 2004, and 2005—the big automakers' fleets would not have met federal fuel-economy standards if not for the credits they got through the FFV scam.[216] As *Consumer Reports* explained in its 2006 report on ethanol, the "FFV surge is being motivated by generous fuel-economy credits" that allow the Detroit automakers to "pump out more gas-guzzling large SUVs and pickups, which is resulting in the consumption of many times more gallons of gasoline than E85 now replaces."[217] In other words, building FFVs allows the Detroit automakers to pretend that they care about fuel economy and, at the same time, to churn out fleets of trucks and SUVs that get lousy gas mileage. So how much fuel has the E85-FFV scam cost the U.S.? In early 2007, *U.S. News & World Report* magazine reported that the U.S. "will burn 17 billion more gallons of gasoline from 2001 to 2008" as a result of the scam.[218]

The implications of the FFV scam go far beyond the ethanol issue and straight to the heart of the competitive abilities of the American automakers. In 2006 alone, GM built some 400,000 FFVs.[219] And GM, along with Ford and DaimlerChrysler, all of which are in lousy financial condition, must continue pumping out big trucks and SUVs because those vehicles are their most profitable models. Meanwhile, Honda and Toyota, both of which are stealing market share from the American automakers, are going their own way. By mid-2007, neither of the Japan-based automakers was producing any vehicles for the U.S. market that were capable of burning E85. Toyota officials say they may begin marketing a pickup truck in the U.S. in 2008 that will be capable

The Ethanol Timeline

1862—The Union Congress places a $2 excise tax on each gallon of ethanol to help pay for the Civil War. Before the war, ethanol was a major illumination oil. After the tax is imposed, its lighting use declines.

1896—Henry Ford's first automobile, the quadricycle, is built to run on 100 percent ethanol.

1906—Congress removes taxes on ethanol.

1908—Ford produces the first Model T, which can run on ethanol or gasoline, or a mix of the two.

1978—The Energy Tax Act of 1978 defines gasohol for the first time: a blend of gasoline with at least 10 percent alcohol by volume. It excludes alcohol made from petroleum, natural gas, or coal and provides a subsidy of $0.40 for each gallon of ethanol blended into gasoline.

1980–1984—Congress enacts a series of tax benefits for ethanol producers and blenders. They include the 1980 Energy Security Act, which offers loans to small ethanol producers. That same year, Congress slaps a tariff on foreign ethanol. Previously, foreign producers, such as Brazil, were able to ship their ethanol (which was cheaper than domestic ethanol) into the U.S. duty-free.

In **1983**, the Surface Transportation Assistance Act increases the ethanol subsidy to $0.50 per gallon. In 1984, the Tax Reform Act increases the subsidy to $0.60.

1990—Congress passes the Omnibus Budget Reconciliation Act, which cuts the ethanol subsidy to $0.54.

1995–1996—Poor crop yields and rising corn prices lead some farm states to pass subsidies to keep the ethanol industry afloat.

1998—Congress extends the ethanol subsidy through 2007, with promises to cut the subsidy from $0.54 to $0.51 by 2005.

2001—Ethanol subsidy cut to $0.53.

2003—Ethanol subsidy cut to $0.52.

2005—Ethanol subsidy cut to $0.51 per gallon.

2007—In his State of the Union address, George W. Bush says the U.S. should increase its production of renewable fuels to 35 billion gallons by 2017.

—Thanks to demand from ethanol plants, corn prices continue to rise.

—Some 75,000 Mexican citizens march in the streets of Mexico City to protest the rising cost of tortillas, which, due to the surge in corn prices, had doubled in price.

Sources: Energy Information Administration, Chicago Board of Trade, Associated Press, the White House, Senator Charles Grassley.[220]

of using E85. But as Bill Reinert, the national manager for Toyota's advanced technology group, told me in April 2007, the company isn't doing it for the CAFE credits. "We don't need the mileage credits," Reinert said.

Toyota and the other big foreign car producers have no need to inflate their fuel economy standards because they are already building more efficient fleets than their American counterparts. In early 2007, the American Council for an Energy-Efficient Economy released its list of the 12 "greenest" cars available in the U.S. The Big Three automakers did not have a single car on the list. Honda and Toyota each had four cars on the list. Nissan and Kia each had one car. Hyundai had two cars on the list.[221] That higher efficiency (along with generally higher-quality products) is having results: In the first quarter of 2007, Toyota passed General Motors as the world's biggest automaker. For the first time since 1931, General Motors is no longer the globe's top producer of cars and trucks.[222]

Despite the many problems faced by the big Detroit-based automakers, they continue to hype FFVs and E85 because corn-based ethanol allows them to continue building massive fleets of big SUVs and trucks that get lousy fuel economy. And they will likely continue doing so until one (or all) of them is forced into bankruptcy.

All of which makes you wonder: What would Lily Tomlin drive?

13

RUNNING OUT OF (NATURAL) GAS

Just for a moment, disregard everything you've read about the ethanol scam. And just for a moment, assume that ethanol and other biofuels have had a huge breakthrough and are magically able to displace the need for imported oil and refined products. In that scenario, America would be energy independent, at least with regard to motor fuels.

While that might be a good thing, the U.S. would still be a big importer of natural gas. In fact, the U.S. has been importing gas for decades, and over the coming decades, those imports are going to rise dramatically. And those imports provide yet another reason why energy independence is bunk.

With everybody talking about "peak oil," the issue of "peak gas" has largely been pushed to the back burner (pardon the pun). But the issue of peak gas has been important since 1973. That year not only marked the global quakes that accompanied the Arab oil embargo, but it also marked the peak of America's natural gas production.[1] And ever since then, American consumers have been relying more and more on Canadian gas. Since 1973, U.S. gas imports from Canada have more than tripled, and Canada has become America's most important gas supplier. In 2002, Canadian gas imports hit an all-time high of some 3.78 trillion cubic feet.[2]

But it appears those days are over. Canadian gas analysts now believe that Canada's gas production peaked in 2002, and they expect that their country's gas output could fall by half over the next two decades.[3] The peaks in production in American and Canadian gas are having a predictable result: More gas imports—in the form of liquefied natural gas (LNG)—will be coming into the U.S. from gas-rich OPEC countries like Qatar and Nigeria.

Between 2000 and 2006, the amount of LNG imports into the U.S. more than doubled, and by 2030, the Energy Information Administration expects natural gas imports, the vast majority of which will be arriving from overseas, to be supplying almost 22 percent of America's natural gas needs.[4]

The Canadians will likely be big gas importers, too. Dave Russum, a technical specialist with AJM Petroleum Consultants in Calgary, expects gas production in Alberta, which produces about 80 percent of Canada's gas, to drop from 14 billion cubic feet per day in 2000 to just 6 billion cubic feet per day by 2024. Even if Canada is able to forestall a major decrease in its natural gas production, the booming tar sands operations in Alberta and Saskatchewan have a huge hunger for gas, which they use to heat the sands to leach out the oil. Some forecasts have even projected that all of the gas that would be carried by the proposed Mackenzie Delta pipeline, which could bring Canadian gas southward from the Beaufort Sea, may be diverted into the tar sands operations.[5]

The crimps in the Canadian gas market have left the U.S. with few options. New LNG-receiving terminals are being built in Texas and on the West Coast to handle the incoming gas. Those terminals will complement the LNG-receiving facilities that are already in place in Maryland, Massachusetts, Georgia, and Louisiana.

While the surging use of imported natural gas may cause some energy isolationists to worry, America's role as a major participant in the global natural gas trade will likely mimic its role as a major oil consumer. In the coming years, tankerloads of natural gas will be plying the oceans just as tankerloads of crude have been doing for decades.

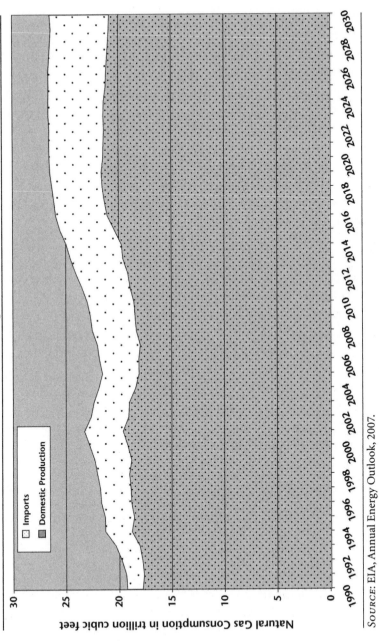

FIGURE 14 U.S. Natural Gas Domestic Production and Imports, 1990–2030

SOURCE: EIA, Annual Energy Outlook, 2007.

And that means that the U.S. will continue buying natural gas from countries like Egypt, Algeria, and Trinidad, all of which will be major gas exporters in the coming decades.[6] New LNG liquefaction plants in Peru may also be supplying gas to the U.S. market.

Other new technologies, including marine shipment of compressed natural gas, will likely speed the development of the global gas trade. Unlike LNG, which requires large amounts of energy to be expended during the cryogenic compression process, compressed natural gas can be transported at lower cost because it doesn't have to be super-cooled. That saves time and money. Energy analysts expect that ships capable of carrying compressed natural gas could begin hitting the global market within five years.[7]

America's ability to produce its own natural gas may be falling, but plenty of other countries around the world have gas to sell. And that gas will find its way to the U.S. gas market, where it will be used to heat homes, power factories, and generate electricity.

Foreign sources of energy play a critical role in America's electricity sector. And few sources are more critical than imported uranium.

14

NUCLEAR POWER

Megawatts from Foreign Uranium

Back in the 1970s and 1980s, environmentalists in Austin, Texas, who opposed the city-owned utility's investment in the nuclear reactors at the South Texas Project had a joke. It went like this:

> Q: What do you have when you have Lassie, Rin Tin Tin, and the South Texas Project?
> A: Two moneymakers and a dog.

The joke was a perfect reflection of those times. In the wake of the Three Mile Island accident in 1979, nuclear power was having a rough time. Plants that were under construction were hampered by cost overruns. Concerned citizen groups were raising questions about safety and nuclear waste disposal. And the South Texas Project, of which the city-owned utility Austin Energy owns 16 percent, was no exception.[1] Approved by Austin voters in 1973, the plant cost far more and took far longer to build than was originally envisioned. The two-reactor power plant was expected to cost $1 billion. It ended up costing about $6 billion. The city's stake was supposed to cost just $161 million. It ended up costing about $1 billion.[2] The plant was supposed to open in 1979 but didn't start generating power until 1988.[3]

In addition to the cost hikes and delays, there were the ever-present worries about safety, particularly due to Three Mile Island and the Chernobyl accident in 1986. Even after the plant began operating, it had difficulties and occasional shutdowns. And in the early 1990s, the city of Austin tried to sell its share of the South Texas Project but couldn't find a buyer. During the height of the battles over the nuclear plant, one prominent Austinite became so convinced of the plant's deficiencies that he declared flatly that it would "never produce economical energy for the city of Austin."[4] (In a curious twist of fate, this former critic of nuclear power is now a top manager at Austin Energy.)

He was wrong. Real wrong. In fact, over the past few years, the South Texas Project has emerged as one of the best deals the city of Austin has ever done. The South Texas Project is one of the most efficient electricity producers in America. For three years in a row, from 2004 to 2006, the plant produced more electricity than any other two-reactor nuclear plant in the country.[5] Austin now gets about 29 percent of its electricity from the nuclear plant, and that juice is likely the cheapest power in its portfolio.

In addition to the nuclear plant, the city also generates power using coal, natural gas, and renewables. But as fossil fuel prices have risen, the value of the South Texas Project has become apparent. In 2001 alone, the city saved about $147 million in avoided natural gas costs thanks to the South Texas Project.[6] Between 2002 and 2006, the city utility's natural-gas fuel costs nearly tripled, going from $82.2 million to $230.4 million per year. Meanwhile, the city's total fuel costs for the South Texas Project in 2006 were just $13.5 million.[7] Thus, between 2001 and 2006, the nuclear plant likely saved the city about $900 million in avoided natural gas costs.[8] In that six-year period alone, the city's share of the South Texas Project likely paid for itself.

While it is true that fuel costs for nuclear plants are a relatively small portion of their overall operating costs, it's clear that the nuclear plant is saving the city big bucks. When counting all of the costs—fuel, operations, maintenance, and borrowing costs—the electricity from the nuclear plant probably costs the city about $0.017 per kilowatt-hour.[9] In 2006, the average cost of power from all of the city's generation sources was $0.039 per kilowatt-hour.[10]

Thus, the old joke about the South Texas Project should be updated: It turns out that Lassie and Rin Tin Tin were the dogs. The South Texas Project is the moneymaker.

The cost figures are only part of the story. With governments and policymakers increasingly concerned about carbon dioxide emissions, the South Texas Project provides the city's only large-scale power source that is either low- or no-carbon. (The city has invested heavily in solar and wind power, but those sources provide only about 6 percent of the city's electricity.[11]) Austin's experience with the South Texas Project demonstrates two key issues regarding the domestic energy business. First, over the past three decades or so, nuclear has been the only energy source that has taken a significant amount of market share away from fossil fuels. Second, the South Texas Project provides another example of the delusion of energy independence. Why? Most of the commercial power reactors in the U.S. use imported uranium.

In the three decades that followed Richard Nixon's first declaration that America would be energy independent, remarkably little changed with regard to the country's overall energy consumption. Yes, cars got more efficient and the economy as a whole became far more efficient in its use of energy. But over those three decades, fossil fuels retained their dominant position, and renewable energy supplies remained nearly constant, at about 6 percent of the country's total energy consumption. The big change in America's energy mix came from nuclear power. During that time period, nuclear power's share of the country's energy mix went from 1 percent to 8 percent.

While critics of nuclear power bring up valid points about its shortcomings, including the need to solve the waste disposal problem and the high cost of building new plants, the undeniable truth is that uranium-based electricity production has had a dramatic effect on the country's energy consumption. The 104 commercial power reactors in America now provide the country with about 20 percent of its electricity.[12] Without the billions of kilowatt-hours that come from

FIGURE 15 Overall U.S. Energy Consumption, by Source, 1973 and 2004

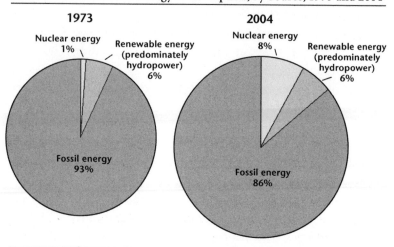

SOURCE: Government Accountability Office, "Department of Energy: Key Challenges Remain for Developing and Deploying Advanced Energy Technologies to Meet Future Needs," December 2006, 9 (GAO-07-106).

those nuclear plants, the U.S. would be using far more coal and natural gas, and possibly even more oil, to generate electricity. And that would mean far greater emissions of air pollutants and carbon dioxide. For all the wailing and gnashing of teeth over the dangers of nuclear power, the U.S. would be in far worse shape from an energy standpoint if it did not have the 104 nuclear reactors that are now pumping out electricity.

Furthermore, a key, but overlooked, aspect of the nuclear power story is the one about imports. And those numbers show the absurdity of the concept of energy independence. Even if the U.S. were somehow able to quit importing oil and quit importing natural gas, it would still be importing lots of uranium.

America's electric utilities are now importing about 83 percent of the uranium they need to power their nuclear reactors.[13] In 2006, American utilities bought uranium from Australia, Brazil, Canada, Kazakhstan,

Namibia, Russia, South Africa, and Uzbekistan. The biggest supplies came from Australia and Russia.[14]

Given America's dependence on foreign uranium for its nuclear power plants, true energy independence for America would require one of two things to happen: (1) The U.S. government would have to spend billions of dollars in upgrading and subsidizing the domestic uranium-mining business so that it could provide for all domestic needs, or (2) Congress would have to mandate the shutdown of the vast majority of the country's nuclear power plants.

From the 1950s to the 1980s, the U.S. was able to provide much of its own uranium from mines in the Rocky Mountains. But the accidents at Three Mile Island and Chernobyl effectively killed all of the planned nuclear power plants that were being discussed at that time. And by killing off the growth of the domestic nuclear power industry, those accidents also strangled America's uranium-mining industry.

The boom in American uranium mining started right after World War II. Between 1949 and 1960, domestic uranium production went from 400,000 pounds per year to 35 million pounds. By 1980, domestic uranium production had hit its all-time high of 44 million pounds. But the plant cancellations took their toll, and by 1985, American production had fallen by 75 percent.[15] Uranium operations throughout the western U.S. were forced out of business. The crash took a particularly hard toll on New Mexico. In 1979, the uranium business was the king of New Mexico's mining business. It employed more miners than any other sector. The town of Grants was the epicenter of 45 mines and five mills. Six years later, all of the mines and mills were closed. More than 7,000 jobs were lost.[16]

Over the past few years, the U.S. uranium industry has been ramping up its production. In 2006, U.S. miners produced 4.7 million pounds of uranium.[17] But even if that production continues growing, it will take years for the domestic production to increase to a level that obviates the need for imports. And it's unlikely to return to the levels seen in 1980. Further, the U.S. simply doesn't have the reserves that other countries do. The U.S. has only about 102,000 tons of uranium

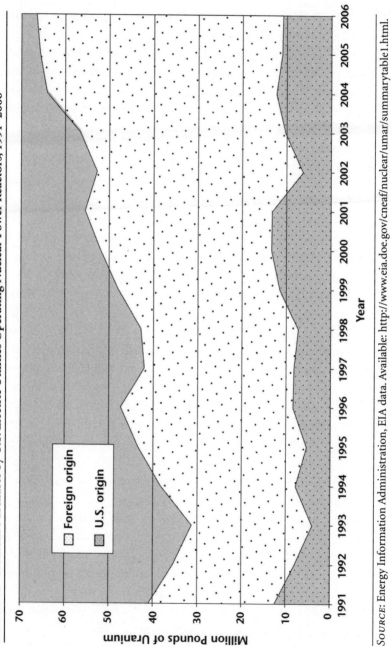

FIGURE 16 Uranium Purchases by U.S. Electric Utilities Operating Nuclear Power Reactors, 1991–2006

Source: Energy Information Administration, EIA data. Available: http://www.eia.doe.gov/cneaf/nuclear/umar/summarytable1.html.

reserves. Australia and Canada each have more than 400,000 tons, and Kazakhstan has over 250,000 tons.[18]

Meanwhile, world demand for uranium continues to grow. More than two dozen new nuclear plants are now under construction around the world.[19] Those plants are located in China, India, Finland, and elsewhere. As those new plants begin operating, they will require uranium.

Several American utility companies are talking about building new reactors. And they are being encouraged by the Bush administration to do so. But if (or when) the U.S. finally decides to increase the amount of nuclear power it uses, those new power plants will likely be producing electricity from foreign uranium. And that will be true even if the U.S. uranium industry continues its recovery.

15

THE "OPIATES
OF ENERGY POLICY"
Coal-to-Liquids and Imported Coal

America is the Saudi Arabia of coal. At current rates of consumption, the U.S. has more than 200 years of reserves left. Those vast coal deposits are spread all around the country, from Appalachia, to the Rockies, to Texas, to Alaska. About 52 percent of all the electricity in the U.S. comes from burning coal. By 2030, that percentage is expected to rise to 57 percent.[1]

Whenever the issue of energy independence is discussed, America's coal reserves are inevitably mentioned as part of the solution to the problem. The argument: If only the U.S. were to tap its coal reserves more effectively, we wouldn't need foreign energy.

But there's a dirty secret about the coal being used in America: The Saudi Arabia of coal is also a coal importer. Although the U.S. is a net coal *exporter*, it also relies on significant imports of coal from a variety of countries, including Colombia, Venezuela, Indonesia, Russia, and even China.[2] Most of these imports are of low-sulfur coal, which is imported by electric utilities that must comply with federal regulations to reduce their sulfur dioxide emissions. And for some utilities, it is cheaper to import that coal than to have it shipped by train from central Appalachia or Wyoming.[3]

Between 2001 and 2006, America's coal imports nearly doubled, reaching some 36.2 million tons in 2006.[4] While that's a relatively small portion of the total annual coal consumption of about 1.1 billion tons, imports are expected to more than double over the next two decades thanks to America's huge thirst for coal.[5] By 2015 or so, America will go from being a net coal exporter to a net coal *importer*. And by 2030, the Energy Information Administration expects that about 10 percent of the coal demand east of the Mississippi River will be supplied by imported coal.[6]

The surge in imported coal is occurring at the same time that there is increasing talk about producing synthetic fuel from domestically produced coal. This approach is being pushed hard by a number of politicians, particularly those from coal-producing states. Montana governor Brian Schweitzer, a Democrat, has been a leading proponent of coal-to-liquids (CTL) technology. He insists that America "can achieve energy independence" if only it increases its energy efficiency, dramatically increases biofuel production, and begins a massive program to turn Montana's vast coal reserves into motor fuel. Further, he claims that the black rocks can be turned into liquid fuel for just $1.20 per gallon.[7]

Other politicos are on the CTL bandwagon. George W. Bush has included CTL in his list of the alternative fuels that he believes will help reduce America's dependence on foreign oil.[8] In 2006, Barack Obama, along with five other senators (four of them were Republicans), introduced legislation called the Coal-To-Liquid Fuel Promotion Act of 2006, to "create a backbone of CTL infrastructure" in the U.S. The bill, which did not pass the Senate, was designed to allow the federal government to extend loan guarantees and tax credits to companies that build CTL plants. "The people I meet in town hall meetings back home would rather fill their cars with fuel made from coal reserves in Southern Illinois than with fuel made from crude reserves in Saudi Arabia," declared Obama. "We already have the technology to do this in a way that's both clean and efficient. What we've been lacking is the political will."[9]

213

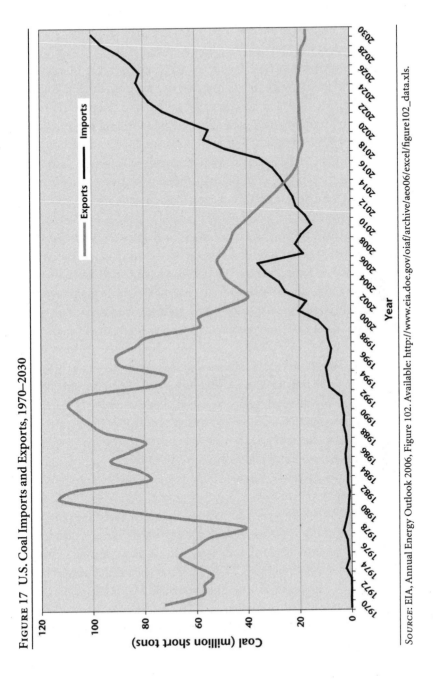

FIGURE 17 U.S. Coal Imports and Exports, 1970–2030

Source: EIA, Annual Energy Outlook 2006, Figure 102. Available: http://www.eia.doe.gov/oiaf/archive/aeo06/excel/figure102_data.xls.

Perhaps. Then again, maybe what's been lacking is profitability.

For six decades, CTL proponents have been promising that huge quantities of motor fuel can be derived from coal and that the maturation of the technology looms just over the horizon. During World War II, the U.S. Interior Department began funding a small CTL program. In 1951, during the Korean War, the Interior Department again flirted with CTL programs and even obtained $455 million in loan guarantees to be made available to entrepreneurs willing to get into the business of turning coal into motor fuel.[10] In 1975, Gerald Ford asked Congress for funding to get CTL plants going. He set a goal of 1 million barrels of synthetic fuel and shale oil production per day by 1985.[11]

In his book *A Policy of Discontent*, Vito Stagliano wrote that coal-to-liquids has "served as the opiates of energy policy most especially in the wake of the 1973 energy crisis." He goes on, saying that CTL has been "touted as the answer to America's energy problems by Republican and Democratic administrations alike." And Stagliano wrote that the economic viability of CTL has always been stubbornly just out of reach:

> In 1950, the difference in cost between synfuels and crude oil was projected to be bridged in five years. Later, it was to be bridged—again in five years—for [Richard] Nixon's Project Independence, and by the same time period for [Jimmy] Carter's National Energy Plan. For George H. W. Bush's National Energy Strategy, the cost difference was projected to be closed by the turn of the century, provided that the real price of crude oil would, on a sustainable basis, be no lower than $29 per barrel.[12]

Lately, CTL proponents are insisting that coal-based fuel can be competitive if oil prices stay above $50 per barrel. But even at that price, CTL plants are enormously expensive. A plant capable of producing just 50,000 barrels of CTL fuel per day will likely cost $4.5 billion.[13] (For comparison, an oil refinery capable of processing 200,000 barrels per day would cost about $5 billion.[14]) The high cost of CTL fuel was confirmed in a March 2007 report by a group of engineers at the Massachusetts Institute of Technology. They estimated that it would cost $70 billion to construct enough CTL plants to replace just 10 percent of the gasoline consumed in the U.S.[15]

In addition to the cost issues, converting coal to motor fuel is a dirty process. CTL plants generally use the Fischer-Tropsch technology, which was developed early in the 1920s by two German scientists, Franz Fischer and Hans Tropsch. The process was used extensively by both the Germans during World War II, in an effort to overcome their chronic motor fuel shortages. South Africa used the Fischer-Tropsch technology extensively during the apartheid era to overcome trade sanctions imposed on it by other countries. The process allowed South Africa to convert its vast coal reserves into motor fuel and chemicals.

Now that apartheid has ended, South Africa's leading energy company, Sasol, is selling its knowledge of the Fischer-Tropsch technique to other countries, including gas-rich Qatar and coal-rich China. The Chinese are in the midst of a $15-billion CTL production program that will make China the world's biggest producer of coal-based motor fuel.[16]

The main environmental problem with the CTL process is the huge amount of air pollution it creates. At Sasol's sprawling Secunda coal mine and CTL production facility in South Africa, a brownish-yellow haze of coal dust and smog can be seen for dozens of miles in all directions. (Sasol's technology is now also being used in China.) In October 2005, I visited the Secunda plant. During that visit, a Sasol official told me that while Chinese visitors are always impressed with Sasol's technology, they always ask the same question: "What do you do with all the air emissions?"

It's not just about smog and particulates. CTL also creates massive amounts of carbon dioxide. In a 2005 speech at the World Petroleum Congress in Johannesburg, an executive for Toyota, Masayuki Sasanouchi, provided the entire "well to wheel" carbon dioxide emissions for a variety of motor fuels. (Well to wheel refers to the life cycle of a fuel, from drilling the well, to refining, to combustion inside the engine of the vehicle.) Sasanouchi used gasoline as the baseline fuel and compared it to diesel fuel, natural gas, liquefied petroleum gas, and other fuels. Of the 23 fuels studied by Sasanouchi and Toyota, motor fuel made from coal had the highest carbon dioxide footprint, releasing about 50 percent more carbon dioxide than gasoline.[17]

Given the many obstacles to profitable CTL production, the Energy Information Administration does not expect it to gain much traction. In its Annual Energy Outlook for 2007, the agency predicted that CTL production in the U.S. would be just 440,000 barrels per day by 2030.[18] If production hits that level, CTL will be supplying less than 2 percent of America's total oil needs by that time.[19]

Unless or until there are major breakthroughs in both CTL production technology and carbon sequestration, it is doubtful that CTL will ever be more than a niche player in the U.S. transportation fuel market.[20]

16

SOLAR

The 1 Percent Solution

During my first year in the business of electric power generation, my solar panels produced 3,861 kilowatt-hours of electricity. During my best months, May and June, the array of 120-watt Kyocera photovoltaic panels on the roof of my house generated more than 400 kilowatt-hours of power per month. During the worst months, December and January, those 27 panels produced 200 kilowatt-hours or less per month.

The output was less than I expected. When I agreed to purchase the system, I had projected that the panels, which have a total rated capacity of 3,240 watts, would produce 4,600 kilowatt-hours of power per year. I also expected my out-of-pocket costs to be about $5,700. Instead, due to my decision to upgrade the inverter and add a couple other items, my total costs were $7,445. That's not bad, as the entire system cost $22,445 and Austin's city-owned utility, Austin Energy, agreed to pay $15,000 of the total bill. The city is pushing solar power in hopes that it will reduce the need for new power plants. Systems like mine, called *grid tie systems,* allow homeowners in urban areas to produce power and feed it into the power grid. This means that on a sunny day, I get to run my electric meter backward.

The panels are now providing about 31 percent of our annual electricity needs. Our consumption is fairly typical. In 2006, my family of five used about 12,300 kilowatt-hours of electricity in our 2,700-square-foot home. (The average American residence is about 2,349 square feet and uses about 10,656 kilowatt-hours per year.[1]) And we pay about $0.10 for each kilowatt-hour that we buy from the city utility, which is also about average for American homeowners.[2] Thus, the solar panels save us about $386 per year. In addition to offsetting some of our power consumption, the panels are an interesting feature on the house. Once a month or so, neighbors and passers-by stop to ask about the panels and their capacity and cost.

Alas, the solar panels have not been a good financial investment. My initial projections were that the system would pay for itself after about 13 years. However, after 2 years of operation, it's clear that recovering my original $7,445 investment will take more than 19 years, or about half again as long as I originally projected. And that payback assumes no cost of capital. If I had borrowed that $7,445, even at a low interest rate, the loan payments would have been nearly double the monthly savings in electricity costs.[3]

On a cash return basis, the panels are yielding the equivalent of about 5.1 percent per year, tax-free. That's not terrible. But I could have done better from a financial standpoint by purchasing stock in an energy company like Exxon Mobil. If, on October 1, 2004, when I bought the panels, I'd used that $7,445 to buy Exxon Mobil stock, which was then selling for about $46.50 per share, by October 1, 2006, those same shares would have been worth about $66.43 each, and the value of my stock holding would have been $10,628, a gain of 42 percent, or more than four times as much as the annual return from the investment in the solar panels. In addition, I would have been getting a dividend check from Exxon Mobil of about 1.5 percent per year. Or I could have bought a high-yielding drug stock like Pfizer or a bond fund, both of which, by mid-2007, were paying dividends of between 4 and 6 percent.[4] I would have taken some price risk on the value of the stock, but I would also have had steady cash income. And I wouldn't have had to worry about the solar panels being damaged by hailstorms or high winds, or about the possibility of a leaking roof.

The point of these numbers is that solar power, despite all of its promise, still has a long way to go before it becomes cost-competitive. The payback on my solar panels is 19 years—and that's with a huge city subsidy. Without that subsidy, the payback would have been about 58 years at current electricity prices, and again, that assumes the money used to purchase the panels is interest-free.

My personal experience with solar power is fairly typical. In April 2007, analysts at the financial services firm Raymond James & Associates released a report estimating that solar power in residential applications costs $0.37 per kilowatt-hour, nearly four times as much as electricity from conventional sources.[5] These high costs, which are largely due to the high cost of manufacturing the photovoltaic panels themselves, will almost certainly prevent solar from being a major player in the American energy mix for years to come. And that means that solar power will not be able to offset the power from large amounts of fossil fuel, imported or domestic.

Furthermore, even if the electricity generated from solar power were able to be used in the automotive sector—the segment of the economy that is most dependent on foreign energy—solar power cannot scale up fast enough to provide significant amounts of electricity. In early 2006, Simmons & Company International, a Houston-based investment banking firm, estimated that even if solar power capacity grew at a rate of 25 percent per year from now until 2020, it would still be able to provide only about 1 percent of global electricity demand by that time.[6] That same study put the total cost of solar power at $0.22–$0.84 per kilowatt-hour.[7]

By 2030, the Energy Information Administration estimates solar power will be providing about 5 billion kilowatt-hours of electricity per year to American consumers. That's a mere fraction of the 3,351 billion kilowatt-hours of electricity the agency expects to be coming from coal. Put another way, by 2030, the U.S. will likely be getting 53 times as much electricity from hydropower as it will be getting from solar. The agency's projections for electricity production in 2030, in billions of kilowatt-hours, are shown in Figure 18.

FIGURE 18 U.S. Energy Information Administration Projections for
Electricity Generation, by Fuel, 2030

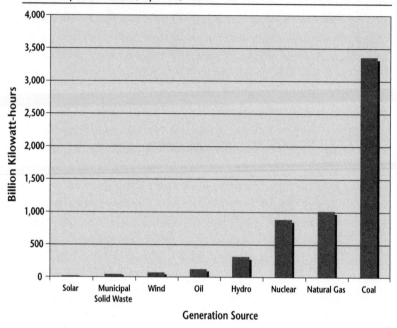

SOURCE: EIA, Annual Energy Outlook 2006, Figures 62, 63, and 64. EIA data.
Available: http://www.eia.doe.gov/oiaf/archive/aeo06/excel/figure62_data.xls,
http://www.eia.doe.gov/oiaf/archive/aeo06/excel/figure63_data.xls,
http://www.eia.doe.gov/oiaf/archive/aeo06/excel/figure64_data.xls.

The other killer drawback of solar power is its intermittency. Lower
power output on cloudy days and during the winter—and zero output
at night—means that solar power facilities must be paired with expensive batteries or conventional power plants in order to prevent blackouts or brownouts. Unless or until there is a significant breakthrough
in high-density electricity storage—a problem that has confounded
scientists for more than a century—solar power can never be relied
upon to provide large quantities of base load power.

In other words, while solar power can provide a fair amount of electricity, it cannot lead America to energy independence. It's too expensive, it cannot provide large amounts of new electric capacity for the foreseeable future, and it cannot displace imported oil because the American automotive fleet runs on liquid fuels, not electricity.

The same problems facing the solar sector are abundant in the wind sector.

17

HOT AIR

Wind power is the electricity sector's equivalent of ethanol: The hype has lost all connection with reality.

Former vice president Al Gore has said that future societies will rely on "diversified and renewable sources of energy, ranging from windmills and solar photovoltaics" to biodiesel and ethanol plants. Gore goes on to claim that these facilities "will make the industrialized world more secure and less dependent on unstable and threatening oil-producing nations."[1]

The environmental group Greenpeace declares that wind power prices are "competitive with the cost of electricity from new coal-fired power plants."[2] Another environmental group, the Natural Resources Defense Council, says on its Web site that wind is "free, inexhaustible and immune from inflation."[3]

In 2000, then-Secretary of Energy Bill Richardson (who's now the governor of New Mexico and a Democratic hopeful for president in 2008) declared that wind power should be providing 5 percent of America's electricity by 2020. Richardson said that increasing the use of wind power will "promote regional economic development" and "increase America's energy security."[4]

The acme of the wind power craze may have been reached in May 2007, when the American Council on Renewable Energy released a report claiming that it is "technically feasible to increase wind capacity

to supply 20 percent of this nation's electricity by 2030."[5] The group
went on to say that the U.S. could have some 340 gigawatts (340,000
megawatts) of installed wind-generation capacity by that time—or
about one-third as much electric generating capacity as the U.S. had in
place in 2007.[6]

But during the council's telephone press conference trumpeting
the new report, I asked a question: How many kilowatt-hours does the
group expect all of those windmills to generate? The answer: They
hadn't done any estimates, and they are "going to continue" working
on it. They might, they said, have estimates of actual output in 2008.

It was a stunning response. It's "technically feasible" to send all of
the residents of Wyoming to live on the moon. But we haven't done
that. It's also technically feasible to install 340 gigawatts of wind turbines, but if only a small portion of those turbines is producing cost-
effective electricity, what's the point?

By the end of 2006, the U.S. had 11.6 gigawatts of installed wind-
power capacity.[7] Although those turbines can produce large amounts
of electricity, they also come with two pivotal limitations that will prevent wind power from providing large quantities of electricity to U.S.
consumers:

1. Like solar power, wind power is intermittent. While that fact is
   obvious, it means that wind cannot replace the need for base
   load electric generating capacity from traditional sources like
   coal, natural gas, and nuclear.
2. Few people want to live near wind turbines.

An example of the problem of intermittency is a city that needs
1,000 megawatts of electric generating capacity. Needless to say, it
needs that 1,000 megawatts to be absolutely reliable so that there are
no unexpected blackouts. To meet the demand, let's assume the city
relies on a mix of coal, natural gas, and nuclear power plants. And in
order to appear more environmentally conscious, the city also decides
to install 100 megawatts of wind turbines. But since the city's power

engineers cannot possibly know when the wind will blow and, thus, when those 100 megawatts of wind power will be available, they cannot forgo any of the 1,000 megawatts of base-load generating capacity with which they began.

On a windy day, the city may be able to use, say, 50 megawatts of power from the wind turbines and therefore dial back some of the output from the conventional power plants. In doing so, the city will save some of the fuel that would ordinarily be burned in those plants. But most, if not all, of those 1,000 megawatts of conventional electric power stations must still be kept running—and ready to supply electricity—just in case the wind stops blowing. In industry parlance, those plants that are kept running at a lower capacity are called the *spinning reserve*—meaning they are there to keep the amount of electricity on the grid from surging or flagging.

The stochastic nature of wind power is particularly problematic during the hottest days of summer, when electricity demand is at its highest. The wind doesn't blow on the hottest days. Therefore, the turbines are not able to supply electric power during times when electricity prices are at their highest levels.

Further, as mentioned in Chapter 16 on solar power, there is no cost-effective method of storing large amounts of electricity. Thus, wind turbines cannot provide reliable power during the periods when demand and costs are highest.

So, while wind power proponents claim that electricity from wind turbines may be cheaper than that produced by fossil-fuel-fired power plants, that doesn't mean the final cost of the electricity delivered to customers is any cheaper. Why? Because keeping those spinning reserves—the ones fired by coal, nuclear, gas, and, in some cases, oil—is very expensive.

Ireland, one of the countries with the greatest potential for wind power development, has found that adding more wind power means higher prices for electricity. A 2004 study by Ireland's electric utility, ESB National Grid, determined that adding significant amounts of wind generation capacity to meet European Union targets for renewable energy "will increase electricity generation costs by 15 percent."[8] The reason is that as the wind turbines provide doses of intermittent power, the existing fossil fuel plants will operate, says the ESB report, "less efficiently and with increasing volatility."[9]

That volatility is due to the intermittent nature of wind power. In July 2006, wind turbines in California produced power at only about 10 percent of their capacity; in Texas, one of the most promising states for wind energy, the windmills produced electricity at about 17 percent of their rated capacity.[10] In early 2007, the Electric Reliability Council of Texas, the independent system operator for the state, determined that just "8.7% of the installed wind capability can be counted as dependable capacity" during peak demand periods.[11]

Studies in Europe have shown similar results. In Britain, a study released in early 2007 found that between October 2006 and February 2007, there were 17 days when the output from the country's windmills was less than 10 percent of their rated capacity, and there were 5 days when output was less than 5 percent.[12] In Denmark, in February 2003, there was an entire week when virtually no wind power was generated by the thousands of turbines located along the country's western coast. In August 2002, the region's wind turbines were actually a net electricity loser, as the steering mechanisms on the turbines took in more electricity than the turbines themselves produced.[13]

The Europeans are also questioning the high cost of wind power, particularly when it comes to reducing carbon dioxide emissions. An early 2005 study in Germany found that it costs up to 77 euros (about 100 U.S. dollars) to avoid emitting 1 ton of carbon dioxide by using wind energy. A study done about the same time by Britain's National Audit Office said wind energy was the most expensive way to cut carbon dioxide emissions in Britain, putting the cost at up to 140 British pounds (about 278 U.S. dollars) per ton of avoided carbon.[14]

Perhaps the most damning study on the cost of wind power was issued in 2004 by Britain's Royal Academy of Engineering. That report determined that when all of the costs of wind power—including the need to maintain all of the base-load generating capacity—are counted, electricity generated by wind costs more than twice as much as that generated by coal, natural gas, or nuclear. The study found that wind power cost 5.4 pence per kilowatt-hour. It determined that gas-fired electricity cost as little as 2.2 pence per kilowatt-hour, nuclear cost 2.3 pence, and coal cost 2.5 pence. Even in a scenario where carbon dioxide emissions are taxed at a rate of 30 British pounds (about 60 U.S. dollars) per ton, the Royal Academy found that wind power would still be more expensive than electricity generated by natural gas, nuclear, or coal.[15]

The second key limitation: Wind turbines are not popular neighbors.

In Germany, rural residents have fought the installations of wind turbines. A member of Germany's parliament said his compatriots are opposing them "because of the disastrous effect on our landscape."[16] In Britain, a nonprofit group called Country Guardian has unified opposition to wind projects. The group says wind power projects will hurt tourism and damage the rural character of many regions. It has issued a manifesto declaring that the British government's effort to use wind to cut carbon dioxide "emissions is misguided, ineffective, and neither environmentally nor socially benign."[17]

Bitter fights over wind turbine installations have occurred in the U.S., ranging from Cape Cod, where the Kennedy family opposes a large offshore wind-power project that would be located near the clan's estate in Hyannisport, to the King Ranch in south Texas.

America's first proposed offshore wind farm, called Cape Wind, has been caught in a protracted battle between the wealthy residents of the Cape and a developer who wants to install 130 turbines in Nantucket Sound.[18] Among the leading opponents of Cape Wind: Robert F. Kennedy, Jr., the scion of the powerful Democratic political dynasty and an outspoken environmental lawyer for the Natural Resources Defense Council.

In a December 16, 2005, op-ed in the *New York Times*, Kennedy argued that Cape Wind was financially feasible only thanks to $241 million in federal and state subsidies. He also claimed that while he supports wind power, "some places should be off limits to any sort of industrial development" and that "our most important wildernesses are those that are closest to our densest population centers, like Nantucket Sound."[19] Kennedy's many critics contended that his opposition was just another example of the NIMBY (not-in-my-backyard) syndrome. Two environmental activists, Ted Norhaus and Michael Shellenberger, quickly penned a response to Kennedy in the *San Francisco Chronicle,* calling his opposition to the wind farm an example of "a worldview born among the privileged patricians of a generation for whom building mansions by the sea was indistinguishable from advocating for the preservation of national parks."[20]

In early 2007, the owners of the fabled King Ranch, one of the largest working ranches in the world, led a push in the Texas legislature to pass a law that would regulate the installation of wind turbines. As Robert Elder, Jr., of the *Austin American-Statesman* reported, the bill was designed to require wind development companies to get permits from the state and to determine whether the noise from wind turbines "interferes with the property rights of nearby landowners." The backlash against wind in Texas is rather unexpected, given that the Lone Star State has been the country's most aggressive installer of new wind capacity. In 2007, the state had 2,780 megawatts of wind generation capacity. If the federal tax incentives now available to wind power developers are extended, that capacity could double by 2015.[21]

The efforts by the owners of the King Ranch to limit wind power installations spawned a minor feud with the neighboring Kenedy Ranch, which at 400,000 acres is less than half as big as the King Ranch. When Jack Hunt, the president of the King Ranch, heard that the managers of the Kenedy Ranch were going to allow the installation of some 240 wind turbines, he charged that the smaller ranch was "sacrificing the long-term value of a rare resource for short-term revenue." He also said the siting of the turbines was "a horrific location."[22]

While advocates and detractors quarrel over the location of wind turbines, which can stand some 400 feet high, wind power advocates purposely conflate wind power with energy independence. The May 2007 report by the American Council on Renewable Energy contains a lovely picture of several wind turbines in front of verdant hills. Beneath it is a striking picture of an array of solar panels against an azure sky. The text immediately adjacent to the photos declares that a "reduction of imported energy provides a more secure future. . . . If we can tap the potential of our domestic renewable energy resources, we can make real progress towards achieving true energy independence."[23]

But just like solar, wind power won't do anything to displace imported oil and refined petroleum products because the American automotive fleet runs on liquid fuels, not on electricity. Nor will these sources make meaningful cuts in the amount of oil used to generate electric power because oil-fired generators provide less than 2 percent of America's electricity.[24]

While the physical constraints on wind power will limit its deployment, wind power developers are happily reaping big subsidies. For the companies that install and operate wind turbines, more than half of the potential return on their investment comes from federal, state, and local tax incentives. In 2006 alone, the federal tax incentives cost some $2.75 billion.[25] And those subsidies could increase dramatically if there is a concerted effort to use wind power on a broader scale.

Yet another problem with wind power is its uneven geographical distribution. (Geographical constraints are also obvious for several other renewable-energy resources like geothermal, tidal power, and solar.) In the U.S., the best locations for windmills are usually far away from the urban areas that need electricity the most. Wind maps show that entire states in the southeastern U.S. have little or no prospects for wind power development.[26] That means that utilizing large amounts of wind power on a truly national basis will require the construction of thousands of miles of new, high-voltage transmission lines. That will mean seizing land through eminent domain, an increasingly controversial and costly process. Those many miles of new power lines will cost vast sums of money and will likely mean higher costs for electricity consumers.

By 2007, the U.S. had more than 20,000 wind turbines installed, and wind power production is expected to continue growing at a rapid rate. But in almost any growth scenario, wind will continue to be a bit player in America's overall electricity portfolio. By 2030, the Energy Information Administration expects wind turbines will be producing about 64.5 billion kilowatt-hours of electricity per year. That will be a fourfold increase over the 2004 wind-electricity production level of 14 billion kilowatt-hours.[27] But those 64.5 billion kilowatt-hours will still be only a rounding error when it comes to America's overall electricity use. By 2030, the EIA expects, Americans will be consuming 5,478 billion kilowatt-hours of electricity per year.[28] That means that wind energy will be supplying a little more than 1 percent of America's total electricity needs by that time.

None of this means that wind power should not be pursued where it makes economic and environmental sense, or that wind energy is bad. But it's also clear that the federal government and state governments have to be more judicious in awarding tax breaks and other incentives to wind energy developers, who are eager to put up as many wind turbines as possible, without regard to the long-term utility of their projects.

———

Whether the issue is ethanol, coal-to-liquids, solar, or wind power, it should, by now, be obvious that the U.S. cannot give up its reliance on oil. Nor can it give up its reliance on coal and natural gas. Put simply, America will be using fossil fuels for many decades to come. And that means that the U.S. will have to continue to deal with the many nations of the world that provide it with those fossil fuels.

Part 4 explores the world's growing interdependence.

Thomas Friedman's Maginot Line

Thomas Friedman is a big deal. He's won the Pulitzer Prize three times. He's a columnist for the *New York Times* and a frequent guest on some of the most prominent TV shows in the U.S. He's written several best-selling books, including his latest, *The World Is Flat* (published in 2005), which was on the *Times* best seller list for more than 100 weeks. *The World Is Flat* was named one of the best business books of 2006, and *U.S. News & World Report* has named Friedman as one of America's "best leaders."

Therefore, when it comes to energy, Friedman must really know what he's talking about. Right?

Alas, if only that were so. Of all the cheerleaders for energy independence, Friedman has likely been the loudest and the most influential. He may also be the least informed. Since 2002, Friedman has used his spot on the *Times* editorial page—arguably the most powerful media pulpit in America—to publish a steady stream of nonsense about the many benefits that America would reap from energy independence. And while doing so, he has regularly contradicted his much-hyped thesis that the world is now "flat."

The contradictions are obvious and abundant.

For instance, consider the thesis of Friedman's book, which is that the flattening of the world means that "we are now connecting all the knowledge centers on the planet together into a single global network, which—if politics and terrorism do not get in the way—could usher in an amazing era of prosperity, innovation, and collaboration, by companies, communities, and individuals."[29] About the time his book was released, Friedman told one journalist that in this "flat" world, money, jobs, and opportunity will "go to the countries with the best infrastructure, the best education system that produces the most educated work force, the most investor-friendly laws, and the best environment."[30]

Friedman sees the process of globalization moving into a new phase, one where individuals can compete on the global level. He calls this "Globalization 3.0" and argues that the world is shrinking "from a size small to a size tiny." In other words, Friedman sees a world in which every person—and every company and every country—will be competing to provide the best, cheapest goods and services to the global marketplace. In this flat world, competition is going to be fierce. And the winners will be the ones who have the resources and the wherewithal to provide the market with the commodities and services at the lowest prices and on the best terms.

continues

Thomas Friedman's Maginot Line *continued*

Unfortunately, the very same Friedman who's seeing the hypercompetitive global market also wants to build a Maginot Line that will protect the U.S. from the Islamic world. In late 2006, Friedman published a column in the *Times* saying that the U.S. should "build a virtual wall. End our oil addiction." Getting rid of our need for oil will, he wrote, "protect us from the worst in the Arab-Muslim world. . . . These regimes will never reform as long as they enjoy windfall oil profits." The solution, he declared, is for America to build "a wall of energy independence" around itself. Doing so "will enable us to continue to engage honestly with the most progressive Arabs and Muslims on a reform agenda."[31]

Friedman's wall ignores the reality of the integrated global energy market—a market in which even the Saudis and Iranians are energy importers. Further, Friedman, a jet-setting journalist who prides himself on his worldly travels and Rolodex of famous friends, appears remarkably uninformed about the lousy history of walls. The Berlin Wall stood for almost 30 years as a symbol of the Cold War until it was undermined by the crumbling Soviet economy and a mass movement of freedom-minded Poles and Germans. During the early months of World War II, the Maginot Line was supposed to protect France from Germany. But the French forgot to tell the Nazis that they weren't allowed to drive their tanks around the wall and through the Ardennes. In 1940, the Germans did just that, the French military quickly crumbled, and, for the next four years, France was a German colony. Going back further in history, the walls of Troy kept the Greeks at bay for a decade or so. But Greek cunning and a very large horse foiled the Trojan fortifications.

If there's one lesson to be drawn from history, it's that walls don't work. And yet Friedman has been claiming for years that a shield of energy independence is the solution to America's woes. In early 2002, he advised President Bush to launch "his version of the race to the moon." Bush should push a "program for energy independence, based on developing renewable resources, domestic production and energy efficiency. Not only would every school kid in America be excited by such a project, but it also would be Mr. Bush's equivalent of Richard Nixon going to China—the Texas oilman weaning America off of its dependence on Middle East oil. That would be a political coup!"[32]

A few months later, Friedman advised Bush that he should create "a Manhattan project for energy independence." And he justified that need for energy independence by using the very same argument that has been made by the neoconservatives, that is, by conflating the issues of oil and terrorism. Energy independence, Friedman declared, would make America "safer by making us independent of countries who share none of our values. It would also have made us safer by giving the world a much stronger reason to support our war on terrorism."[33] In 2004, Friedman reiterated his plea for a "Manhattan project for energy independence and conservation," so that America can break its "addiction to crude oil."[34]

Friedman's lack of savvy about the global energy trade was even more obvious in an August 2005 column in which he again conflated the issues of oil and terrorism ("We are financing both sides in the war on terrorism: our soldiers and the fascist terrorists," he wrote) while also insisting that many of the technologies needed for energy independence are "already here—from hybrid engines to ethanol." He then quoted Gal Luft, the neoconservative who heads the Institute for the Analysis of Global Security and launched the Set America Free campaign. Luft told Friedman that Brazil's success in cutting its oil imports was due to the South American country's "bringing hydrocarbons and carbohydrates to live happily together in the same fuel tank." In Luft's view, ethanol has brought "Brazil close to energy independence" and insulated it from higher oil prices.

Neither Friedman nor Luft bothered to mention the pivotal role that Brazil's national oil company, Petrobras, has had in Brazil's energy mix.[35] In fact, Friedman's story ignores Petrobras altogether, a startling omission given that in 2005, the company was producing about 1.7 million barrels of oil per day. That 1.7 million barrels of oil contained roughly eight times as much energy as all of the ethanol that Brazil was producing on a daily basis in 2005, a fact that escaped Friedman's attention.[36]

Friedman's lack of knowledge about the energy business is also obvious in *The World Is Flat* (in which he also dutifully quotes Luft). In a section devoted to energy issues, Friedman says that the U.S. should be using more electricity in the transportation sector so as to reduce oil consumption. He then states, "We don't import electricity."[37] Wrong. The U.S. has been importing electricity from both Canada and Mexico for many years and will likely continue doing so for many years to come.[38]

Friedman's problem is that he wants it both ways: He espouses the merits and potential of the new flat world, while also insisting that

continues

Thomas Friedman's Maginot Line *continued*

the U.S. should withdraw into energy isolationism and thereby sur-
render any participation in the world's single biggest business. The ir-
reconcilable contradictions in Friedman's arguments are easily seen in
the penultimate paragraph in *The World Is Flat,* where he states the
two greatest dangers he thinks Americans face. The first is "an excess
of protectionism—excessive fears of another 9/11 that prompt us to
wall ourselves in." The other danger, claims Friedman, is fear of competing in the world that "prompt us to wall ourselves off, in search of
economic security." Both stances, he wrote, "would be a disaster for
us and for the world."

So, to summarize Friedman's worldview, he wants a "wall of energy
independence" around America while simultaneously warning Americans that the two greatest dangers are (1) walling "ourselves in" and
(2) walling "ourselves off."

Friedman sees a flat world where walls are dangerous because they
will isolate the U.S. from other countries. But when it comes to energy,
walls are good because they isolate the U.S. from other countries.

Is anyone else here confused?

THE WORLD OF INTERDEPENDENCE

18

THE VIEW FROM JUBAIL

The Chinese never ask us about price. They only want to know about supply.

PRINCE SAUD, CHAIRMAN,
SAUDI BASIC INDUSTRIES CORP., 2006

The scale of the construction projects under way in Jubail, Saudi Arabia, boggle the mind.

Three decades ago, Jubail, located in the oil-rich Eastern Province, was just a dusty desert village. Today, about $200 billion worth of investment later, the city contains one of the largest concentrations of industrial plants on the planet as well as a modern city, shopping center, and huge port. And those industrial plants produce a huge variety of petrochemicals, as well as motor fuel, plastics, steel, and other materials that are shipped to more than 100 countries around the world. The industrial plants at Jubail cover some 1,000 square kilometers, and the Saudis are pushing yet further into the desert. The next industrial park is called Jubail 2. When it is finished, Jubail 2 will cover another 1,000 square kilometers of territory.

If there's any better example of the growing world of energy interdependence than Jubail, it's hard to imagine what it might be. Over the past few decades, the Saudis have been looking at the future of the global energy market, and they have decided that they want to sell as many commodities as they can. In the past, they sold crude. In the future, they want to sell products that have energy embedded in them, like steel, fertilizer, and other products that require the consumption of large amounts of fossil fuels.

To get ready for Jubail 2, the Saudis (backed by an army of laborers imported from Pakistan, Bangladesh, and other countries) are hauling millions of tons of sand, digging miles of canals, and constructing dozens of miles of roads and power lines. Those workers, backed by vast fleets of trucks and heavy equipment, are leveling and preparing mile after mile of barren sand west of Jubail in order to make space for the new industrial facilities. And while the endless lines of dump trucks carrying sand from one location to another are impressive, it's the seawater pipes that really show the project's scale. The individual sections of pipe are cartoonishly large: about 4 meters in diameter and 15 meters long. Stitched together in long parallel lines, they stretch into the distance as far as the eye can see. The pipes will carry water from the Persian Gulf inland for about 10 miles so that it can be used as cooling water by the industrial plants at Jubail 2.

Jubail 2 won't be the end of the Saudis' efforts. They are already discussing Jubail 3 and Jubail 4. "The challenge of the future is how to create more industry out of crude," said Abdullah Saad al-Rabeeah, the president of the Arabian Petrochemical Company.[1] During my March 2006 visit to the sprawling ethylene plant in Jubail that he runs, al-Rabeeah made it clear that the Saudis learned their lessons from the boom-and-bust cycle that gripped the crude oil business during the 1970s and 1980s. They don't want to be dependent on the vagaries of the crude oil market. Selling crude oil is good. But selling higher-margin products like gasoline, diesel fuel, polyethylene, and other more valuable commodities is better—and far more profitable.

The Saudis won't have any trouble getting those products to market. Jubail's massive port reaches deep into the light blue waters of the Persian Gulf. That port allows the world's biggest oil tankers and cargo ships to load and offload their commodities. (Jubail was one of the main logistics and operations bases in the Persian Gulf for the U.S. military during the First Iraq War.) Like everything else at Jubail, the

Cooling water pipes at Jubail. Early 2006: These pipes will carry seawater to industrial plants being built at Jubail 2, a massive petrochemical and industrial zone in eastern Saudi Arabia. The project is one of the biggest construction projects on earth. Investments at Jubail 2 will total $100 billion or more.
PHOTO CREDIT: Royal Commission for Jubail and Yanbu.

port is being expanded, too, in a project that will cost another $770 million or so.

During my visit to Jubail, I met Prince Saud, the chairman of Saudi Basic Industries Corporation.[2] SABIC, which had $20.8 billion in revenues in 2006, is one of the world's largest petrochemical firms.[3] It's also the single largest and most valuable company traded on the Saudi stock exchange, the Tadawul. Shortly after we sat down, Saud asked me about George W. Bush's State of the Union speech, which had occurred just a few weeks earlier. During that speech, Bush had said the U.S. was "addicted" to oil and that America should "make our dependence on Middle Eastern oil a thing of the past."[4]

After a short bit of discussion about Bush and America's oil de-
mand, Saud told me that Saudi Arabia is not overly concerned about
America's oil consumption. His country, he said, is looking toward
China, India, and the rest of Asia, where oil demand is soaring. In fact,
Saud had just returned from China.

"You know, the Chinese never ask us about price," he told me. "They
only want to know about supply."

It was a telling comment. For the Chinese, it's the availability of
crude, not the price, that matters. When the time comes to pay the bill,
the Chinese will pay it. They recognize the central importance that oil
has in their economy. For the Saudis, the Chinese are just one customer among many. The surge in oil prices, combined with the surge
in global demand for commodities of all types, has allowed the Saudis
to invest huge sums of money in new industrial capacity that will allow them to solidify and diversify their economy and create jobs for
their growing population.

Further evidence of the Saudis' desire to move further into manufacturing came in May 2007, when General Electric announced that it was
selling its huge global plastics division to SABIC for $11.6 billion. The
deal was the biggest acquisition by an Arab company of a foreign rival. It
also means that SABIC will see its revenues grow by about one-third.[5]

The sprawling industrial growth under way at Jubail and SABIC's
purchase of GE's plastics division provide just two examples of the ongoing emergence of Saudi Arabia—a country that sees its future as a
hub that provides both energy and manufactured products to both the
West and the East. That strategy became clear when I visited the Saudi
Oil Ministry.

———

The interior of the Saudi oil ministry building in Riyadh looks like
a movie set. The elevators are round, have curved doors, and travel
inside big gray metal tubes. It's as quiet as a marble-plated morgue.
Few people are walking about inside, and even fewer outside. The
building itself is one of the most heavily guarded in the city. It's sur-

rounded by a thick, tall metal fence. The exterior of the oil ministry is rather plain-looking, particularly in a city like Riyadh, which is dominated by two spectacular examples of Space Age architecture: the al-Faisaliyah Center and the 992-foot-high Kingdom Centre, which was built by the Saudi billionaire Prince Alwaleed bin Talal.[6]

The entrance to the oil ministry is a narrow driveway, defined by thick concrete barricades, that's barely wide enough for a full-size automobile. The barricades form a half circle around the perimeter of the building and funnel cars to a checkpoint, where armed guards inspect the credentials of everyone entering the parking lot. Directly in front of that checkpoint is a military truck manned by two soldiers who stand near a .50-caliber machine gun mounted in the bed of the truck.

The security measures were understandable. Saudi Arabia's oil sector has been repeatedly targeted by al-Qaeda. In fact, the day before my early 2006 visit to the ministry building, security forces had thwarted an attack on the Saudis' single most important energy asset: the massive processing plant at Abqaiq.

None of that strife was evident when I was talking to Ibrahim al-Muhanna, a top adviser to the Saudi oil minister, Ali al-Naimi, and to Prince Abdulaziz bin Salman, the Saudi royal family's representative within the ministry. For more than 25 years, al-Muhanna has been a leading player in Saudi oil politics, and he's respected throughout the global industry as a thinker and strategist. Shortly after I sat down in his office, which has windows overlooking central Riyadh, the subject turned to oil embargoes and the future of Saudi-American relations. Al-Muhanna, who looks to be in his early 60s, said the Saudis and the other oil exporters in the Persian Gulf are not interested in shutting off the flow of oil to the West. Doing so would hurt the Saudis as much as, or more than, it would hurt the West, al-Muhanna told me. Plus, he points out that the 1973 embargo didn't really work. "The oil exporters cheated," he said, by selling cargoes to middlemen who then shipped the crude on to the U.S.

Al-Muhanna then turned to a discussion of oil prices and the issue of global energy security. The concept of energy security, he declared, is quite simple. It means, he said, that "you have spare capacity in crude oil production and refining."

FIGURE 19 OPEC Downstream Expansion Plans, 2006–2011

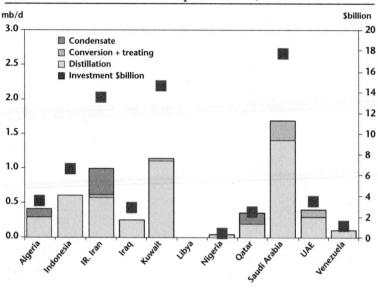

SOURCE: Secretariat of the Organization of the Petroleum Exporting Countries.

Right now, the Saudis are one of the few countries on earth with spare crude production capacity. And they are working hard to increase their refining capacity. By 2011, the country will have invested about $18 billion to add another 1.5 million barrels per day of new refining capacity at locations like Ras Tanura, Yanbu, Jubail, and Rabigh.[7] The new plants, said al-Muhanna, will be among the most advanced in the world. And they will give the Saudis even greater energy security at a time when the U.S.—along with most other industrialized countries on the planet—have neither excess crude capacity nor excess refining capacity.

The punch line here is obvious: Like it or not, America's energy security is tied to Saudi Arabia's energy security. And that's going to be true for years to come. The Saudis are aggressively embracing the global energy market. But try as they might, they are going to have difficulty keeping up with the rock-and-roll capitalism and go-go globalization that are under way in Dubai.

19

THE RISE OF DUBAI AND
THE "SHIFT IN GRAVITY"

It was a forest of construction cranes. On both sides of Sheikh Zayed Road, the highway that connects downtown Dubai with the massive port and free trade zone at Jebel Ali, the cranes were at work, building offices, apartment buildings, condominiums, shopping malls, and warehouses.

In mid-2006, the in-flight magazine for Emirates Airlines, the state-owned carrier, claimed that one-quarter of the world's construction cranes were working in Dubai. And a trip down Sheikh Zayed Road appeared to confirm that assessment: The entire region was one big, hot, dusty construction zone. Skyscrapers were going up by the dozens—each with even more exotic geometry than its neighbor. There were other indicators of Dubai's emergence as a world power. By mid-2006, office rents in the emirate were among the highest in the world—vying with London and Tokyo for that distinction.[1] Shipping volumes through the huge port at Jebel Ali, the largest human-made harbor between Singapore and Rotterdam, continue to soar. And feeding all of that growth is an enormous amount of financial activity.

Dubai has become a transportation hub, a financial hub, and, of course, an energy hub. Dubai is one of the seven sheikhdoms that make up the United Arab Emirates, a country with some 97 billion barrels of oil reserves.[2] But even with all that oil, the UAE provides a couple of examples of the fundamental reality of global energy interdependence.

The first example comes from the country's trade figures. In 2005, the UAE exported an average of nearly 2.2 million barrels of crude oil per day.[3] But that year it also *imported* an average of 259,000 barrels of gasoline and other oil products per day.[4] In fact, of all the Arab oil exporters, the UAE is among the biggest importers of refined products.

The second example: On June 1, 2007, the Dubai Mercantile Exchange began trading both crude oil futures and jet fuel futures. Within a few days of its launch, the exchange was trading several thousand contracts per day. To outsiders, this may appear to be a minor event, but the rise of a new energy-trading hub in the Middle East is indicative of a broader shift in financial power away from New York and London, and toward the Persian Gulf.

Of course, Dubai is unlikely to bankrupt the financial powerhouses on Wall Street and in The City anytime soon. But the creation of the Dubai Mercantile Exchange shows that the global energy market is becoming more regionally diverse, with increasing amounts of capital flowing through the banks and financial institutions in the Persian Gulf. By getting into the futures business, Dubai is ensuring that the crude oil coming out of the Persian Gulf will have its own benchmark price—one that is not reliant on Western crude oil standards like those of West Texas Intermediate and North Sea Brent. The new crude futures—called the Oman Crude Oil Futures Contract—will likely give more pricing power to the oil producers in the Persian Gulf, and it will allow them to hedge their price risk. In July 2006, over breakfast at the Emirates Towers Hotel, Gary King, the CEO of the Dubai Mercantile Exchange, told me that the emergence of the exchange and the new futures contract indicates that the Persian Gulf is "the center of the world's biggest hydrocarbon province. Most of the growth in oil consumption is in Asia-Pacific. So it's a natural shift in gravity. Our timing is very opportune to be in that center of gravity."

King's point is well taken: There's an ongoing shift in the globe's center of gravity—and that gravity is moving away from the U.S. and Europe and toward the Persian Gulf and Asia. Dubai is a prime example of that shift. Whether the issue is finance, oil production, trade, or shipping, the little emirate on the Persian Gulf (thanks in no small part to an army of imported workers from southern Asia) has become a powerhouse.[5] The tiny emirate covers just 1,588 square miles (about the size of Rhode Island), and yet it is constructing the Burj Dubai, which, when completed in 2008, will be the world's tallest building, at some 2,300 feet high.

Dubai provides proof that the global marketplace—in crude oil futures, iPods, fresh flowers, cell phones, beer, shoes, TVs, and a million other items—is becoming ever more integrated. And that's a key point when it comes to energy: As the world's markets become more financially interconnected, the key factors in determining strategic interests are increasingly economic interests, not military ones. The boom in Dubai is perhaps the best indicator of that integration. Dubai has become an economic power by opening its borders and opening its markets. And that openness includes the trading of crude and jet fuel.

The ongoing trend toward global energy interdependence can be seen in the March 2007 decision by Halliburton to move its top executives to Dubai. That move was immediately attacked by Vermont senator Patrick Leahy, a Democrat, who said the move was designed to dodge taxes and was an example of "corporate greed at its worst."[6] Hillary Clinton declared that it was "disgraceful."[7]

While Halliburton's move drew the ire of some American politicos, the decision is another example of the ongoing shift in the global financial and energy markets toward the Persian Gulf. Halliburton recognizes that its future business prospects will hinge largely on its ability to do business with the national oil companies, not with the Chevrons and Exxon Mobils of the world. That point was made clear in the company's March 11 press release, which said that it was making the move to "focus on expanding its customer relations with national oil companies."[8] And Halliburton's CEO, David Lesar, said he "will spend the majority" of his time in Dubai.

While the Democrats may continue howling about Halliburton's move (the company says it will still be domiciled in the U.S. and continue paying American taxes), the decision to station its top officials in Dubai makes economic sense. In 2008, the company expects about 55 percent of its services business to come from the Eastern Hemisphere. That's a big increase over the 40 percent or so it derived from that region in 2006.[9]

The plain truth is that Halliburton's move to Dubai is merely a reflection of the fact that energy companies go where the energy is. Proof of that can be seen in America's failed effort to isolate Iran.

20

ISOLATE IRAN?
DON'T COUNT ON IT

Over the past couple of years, Iranian president Mahmoud Ahmadinejad has become the energy business's equivalent of Monty Hall. While Ahmadinejad has riled the U.S. and Israel with his combative talk and Iran continues to pursue a nuclear development program, the country's state-owned oil company, the National Iranian Oil Company, and the state-owned gas company, the National Iranian Gas Company, have become the modern incarnations of Hall's old game show *Let's Make a Deal.*

In the first half of 2007 alone, the Iranians cut a number of energy deals that will alter the energy balance of power in Asia and the Middle East. Those deals—worth perhaps $50 billion over their lifetimes—show that America's effort to isolate Iran has failed. The world is too hungry for energy, and Iran's energy riches are too vast, for that isolation effort to work. In fact, Iran—perhaps more than any other major energy producer on the planet—provides the quintessential example of energy interdependence.

That energy interdependence comes thanks to a series of contradictions that are peculiar to Iran, a rapidly growing country where energy demand is soaring. Iran has a surfeit of crude oil deposits. It also has staggering quantities of natural gas. At current rates of extraction,

Iran's natural gas reserves—some 970 trillion cubic feet—could last 300 years or more.[1] Iran holds about 15 percent of the world's gas and is second only to Russia in the quantity of gas reserves within its borders.[2] But even with those massive reserves, Iran cannot supply all of its own motor fuel needs and must import about 40 percent of its gasoline, at a cost to the government of about $5 billion per year. Thanks to huge government subsidies, Iranian motorists enjoy some of the cheapest gasoline on the planet, paying about $0.09 per liter. That supercheap gasoline has spawned a wave of smuggling by black market operators, who purchase the subsidized fuel and then ship it to neighboring countries where they can get a higher price.[3]

While Iran struggles to keep its citizens supplied with motor fuel, foreign investors are flocking to Tehran. On January 7, 2007, the National Iranian Oil Company announced the signing of a $16-billion deal with Malaysia's SKS Ventures to develop the Golshan and Ferdos gas fields and build plants to produce liquefied natural gas. The 25-year deal will allow Iran to exploit the two fields, which together contain about 60 trillion cubic feet of gas.[4] Just for reference, the gas in those two fields is about equal to one-third of all the natural gas reserves in the U.S.[5]

On January 26, 2007, Iranian officials announced that they had agreed with representatives from Pakistan and India on the pricing formulas for the long-discussed, much-delayed Peace Pipeline, the $7-billion, 1,600-mile pipeline that will carry Iranian gas to Pakistan and India. (The Bush administration, which continues to hope that it can isolate the Iranian government, opposes the pipeline deal.) Although the deal has not been finalized, the announcement of the pricing breakthrough lends credence to the belief that the pipeline was inevitable. In 2005, Susil Chandra Tripathi, the secretary of India's ministry of petroleum and natural gas, promised that the deal would go through. He told me that the U.S. may "want to isolate Iran, but that doesn't mean Iran will quit producing crude oil and gas, or that we will stop buying it."[6]

On January 29, 2007, Shell and Spain's Repsol-YPF announced that they will help develop Iran's South Pars field in a deal that will likely be worth some $10 billion. (The South Pars is contiguous with Qatar's

North Field. Together, they form one of the world's largest gas fields.) Shell authorities were skittish about disclosing too many details of the deal for fear that it would upset the Bush administration, which is trying to close off foreign investment in Iran. That reluctance was acknowledged by Shell's CEO, Jeroen van der Veer, who said that the company has "quite a dilemma" about its investment.[7]

The January 2007 gas announcements came on the heels of a string of multi-billion-dollar deals between the Iranians and the Chinese. In December 2006, the Chinese National Offshore Oil Corporation signed a $16-billion deal with the Iranians to develop the northern section of the Pars gas field.[8] That deal followed a 2004 gas deal between the Chinese supermajor Sinopec and the Iranians that was valued at $100 billion. But Sinopec's involvement in Iran may be far larger. Sinopec will also take the lead in developing the huge Yadavaran oil field in southern Iran. The field is the biggest onshore field in Iran and contains some 3 billion barrels of recoverable reserves. In all, Sinopec's total investments in Iran may reach $120 billion.[9]

That's just the Chinese. The Indians are also pushing into Iran in a big way. In addition to the huge gas pipeline deal, the Indians are looking to create refining deals with the Iranians. And in 2005, ONGC, India's state-owned oil company, agreed on a $40-billion deal to import liquefied natural gas (LNG) and develop oil fields in Iran.[10]

Don't forget the Brazilians. In March 2007, Petrobras signed a $470-million deal with the Iranians to explore for oil in the Caspian Sea. The Iranians are eager to capitalize on the Brazilians' prowess in deepwater exploration. And Brazil is so eager to get into Iran that it will operate as a service provider, without taking an equity stake in the projects it drills. Instead, it will be paid a share of any oil sold from the fields it develops.[11] Petrobras is also drilling in Iranian waters in the Persian Gulf. The company is working with Repsol-YPF on an exploration project in the Tosan block, a 6,300-square-kilometer region near the Strait of Hormuz.

Brazil plans to continue investing in Iran. During a March 31, 2007, visit to Camp David, Brazilian president Luiz Inacio Lula da Silva was asked at a press conference about Petrobras's investments in Iran. Lula responded, "Petrobras will continue to invest in oil prospection [*sic*] in Iran. Iran has been an important trade partner for Brazil. They buy from us more than $1 billion" in goods and services per year. He continued, "I know that there's political divergence on this between Iran and other countries, but with Brazil, we have no political divergence with them, so we will continue to work together with Iran." For his part, George W. Bush, not wanting to ruffle any feathers, refused to criticize Brazil or Petrobras. Bush said only that "every nation makes the decisions that they think is [*sic*] best in their interest. Brazil is a sovereign nation; he just articulated a sovereign decision."[12]

Don't forget the Austrians. In April 2007, Austria's OMV, the largest oil company in central Europe, signed a deal to develop part of the South Pars field. The deal also includes an agreement for the Austrian company to buy LNG from Iran. News of the deal led a U.S. State Department spokesperson to scold the Austrians, saying that this "isn't really the right time to be thinking about making large investments in the oil and gas sector and, in effect, supporting this regime."[13]

Don't forget the Turks. In September 2007, Turkey's state-owned oil company, Turkiye Petrolleri, announced that it would invest $3 billion over 10 years on a gas project in Iran's South Pars gas field. In addition, the Turkish pipeline company, Botas, has said it will build a pipeline to carry the gas from South Pars to Turkey.[14]

The spate of deals being done with the Iranians exposes the failure of America's attempt to isolate the energy-rich country. Since about 1979, the second-largest oil producer in OPEC has been shut out of the U.S., the biggest oil market on earth, and yet it has never had trouble selling its oil. Today, Iran is exporting about 2.5 million barrels of crude per day. And it is selling every barrel it offers for sale—even though none of that crude flows into refineries in the U.S. Instead, it flows to refineries in countries that are among America's most important allies. The biggest buyer is Japan, which purchases more than 500,000 barrels of Iranian crude every day. Other big buyers are South Korea, Italy, France, and the Netherlands.[15]

There are plenty of other ways to show how America has failed to isolate Iran. Look at Halliburton. It worked in Iran for years through subsidiaries incorporated in the Cayman Islands.[16]

While he was the head of Halliburton, Dick Cheney repeatedly lobbied the federal government to repeal or waive the Iran-Libya Sanctions Act, a law that prohibits American companies from doing substantial amounts of trade with those two countries. After Cheney became vice president, Halliburton was investigated by the government for possible violations of the law. In early 2007, the Halliburton media affairs office refused to discuss the Iranian contracts, and it provided the following statement: "Halliburton intends to wind up its work in Iran and not enter into any other future contracts, and will exit upon the completion of existing commitments."[17] On April 9, 2007, the company issued a brief press release that said it was "no longer working in Iran."[18]

Regardless of whether the company is Shell or Halliburton or Petrobras, it's abundantly clear that the energy business will find ways to work around any sanctions imposed on Iran—regardless of who puts them in place. The companies are doing what they get paid to do: developing energy resources.

The world's hunger for energy cannot be stopped. Iran's vast oil and gas resources are going to be developed and sold to the highest bidder—regardless of what America wants. Nuclear program or not, Iran is emerging as a far stronger regional force in the Persian Gulf, and no amount of posturing by the U.S. will change that fact. Indeed, for all of the Bush administration's piety about the need to isolate Iran, the outcome of Bush's policies appears to be exactly the opposite of what was intended: The one being isolated on the energy front isn't Iran; it's the U.S.

21

A FEW SUGGESTIONS

When it comes to energy issues, Americans are far more interested in rhetoric than reality.

For proof, look no further than the massive energy bill passed in August 2007 by the U.S. House of Representatives. Upon its passage, House Speaker Nancy Pelosi lauded the 780-page bill, known as "A New Direction for Energy Independence, National Security, and Consumer Protection," declaring, "Energy independence is a national security issue, an environmental and health issue, an economic issue, and a moral issue." (What? No mention of male-pattern baldness?)

The bill does nothing to increase energy supplies. It won't cut oil imports, it won't strengthen America's creaky electricity grid, nor will it make America energy independent. Other than that, it's great legislation.

New ways of thinking about energy are needed. So what would I do if I were America's energy czar? Here are a few ideas.

GET GOVERNMENT THE HELL OUT OF THE ENERGY BUSINESS

What does that mean? It means recognizing that federal and state government interventions usually make energy more scarce and more expensive. It also means that government should eliminate *all subsidies and price controls on energy.*

For too many decades, Congress has tried to pick the fuels or technologies that it thinks should prevail. And that effort to pick winners has led to numerous costly programs that have done nothing to reduce prices or increase supplies.

The best example: the ethanol scam. By politicizing the issue of imported oil and hiding behind the flag of "energy security," the farm lobby has created a multi-billion-dollar subsidy machine for the ethanol industry. The results, as detailed earlier, have been higher food prices, increased water use, and dirtier air. In return, American voters have been rewarded with relatively small amounts of corn ethanol, which, when analyzed on a per-gallon cost basis, are far more expensive than conventional gasoline.

Or consider the history of the federal efforts to create synthetic fuel from coal or oil shale, a history that extends back to World War II. The biggest push in this sector was the 1980s-era Synthetic Fuels Corporation, which, like all the previous efforts, was justified by the argument for energy security. And yet, despite the many billions of taxpayer dollars invested in these programs, not a single gallon of cost-effective motor fuel has ever been produced.

Consider natural gas price controls. Throughout the 1970s and 1980s, federal price controls on gas kept supplies tight and prices artificially high. After the natural gas market was deregulated in 1992, gas utilization increased, and consumers benefited by getting cheaper electricity and cleaner air. In the deregulated environment, gas could compete more effectively with coal for market share in electric power generation. In the decade or so prior to deregulation, the amount of coal-fired electricity generated in America jumped by nearly 40 percent, while growth in the amount of natural-gas-fired electricity was minimal. In the years following deregulation, natural-gas-fired electricity has predominated. Between 1992 and 2008, the amount of electricity generated with gas jumped by nearly 87 percent. Over that same time period, coal-fired electricity production increased by 31 percent.[1] Consumers benefited from the deregulated gas business, as electric utilities and other electricity producers were able to build lower-cost gas-fired power plants that could produce electricity more cheaply than comparable coal plants. And those new gas-fired plants emitted far fewer pollutants (like sulfur dioxide and mercury) than coal-fired plants.

The gasoline price controls of the 1970s are another prime example. In the wake of the 1973 oil embargo, gas lines in the U.S. were not caused by shortages of crude; they were caused by price controls that limited the oil companies' ability to sell their refined products for a profit. Predictably, those price controls led to a shortage of gasoline and the infamous queues of motorists waiting for fuel.

Or consider the myriad blends of gasoline now mandated across the country. Nearly four dozen different blends of gasoline are now being sold in the U.S. because of local mandates aimed at improving air quality. These mandates, according to a 2005 report issued by the Government Accountability Office, have "put stress on the gasoline supply system and raised costs, affecting operations at refineries, pipelines, and storage terminals."[2]

While reducing energy regulations, state and federal governments should also eliminate subsidies. There is no need for the federal government to be giving tax breaks to Big Oil for offshore drilling. Given current crude prices, the oil industry has plenty of incentive to drill. Nor is there any need for federal subsidies for ethanol, or for wind or solar power. If the government wants to encourage renewable energy production, that's fine. Offer incentives that encourage low- or no-carbon energy sources, but don't give preferential treatment to any of them. If the government wants to provide improved energy efficiency, it should set benchmarks and let private industry and the marketplace decide which technologies should prevail.

None of these points is made to argue against all regulation. Federal and state authorities must be robust in their efforts to protect human health, in particular. But energy legislation should be guided by the answers to a few questions: (1) Will it increase supplies? (2) Will it lower prices? And (3) does it protect human health, private property rights, and the environment?

If the answers are all affirmative, the legislation deserves consideration.

STOP OBSESSING OVER PRICES AND
REDUCE THE NUMBER OF FUEL BLENDS

Americans are obsessed by motor fuel prices. Over the past three or four years, there has been a barrage of stories on TV, in newspapers, and on the Internet about the rising cost of gasoline. But the truth is that prices fluctuate. They go up and they go down. Get over it.

Americans obsess over fuel prices even though the real cost of motor fuel is flat or declining. While it's true that the nominal price of gasoline has risen significantly over the past few years, when adjusted for inflation the price of gasoline in 2007 was about the same as it was back in the 1930s and early 1940s, and substantially less than it was back in the 1920s. From 1919 through about 1922, the average cost of gasoline in the U.S., adjusted for inflation, was about $3 per gallon.[3] Some nine decades later, in August 2007, American motorists were paying just $2.78 for a gallon of regular gasoline.[4]

By nearly any measure, gasoline is a bargain. Although consumers complain ad nauseum about the grave injustice of paying $3 or so for a gallon of gasoline, they happily troop to Starbucks in the morning and plop down the equivalent of $22.72 per gallon for a venti latte. And they cheerily guzzle Budweiser beer while paying the equivalent of $10.65 per gallon.[5]

No matter how you slice it, gasoline is cheap, cheap, cheap.

Furthermore, the cost of gasoline—when measured as a percentage of the total cost of car ownership—is lower today than it was in the 1970s and 1980s. In 1975, gasoline made up 33.4 percent of the total cost of owning a car. By 2005, gasoline costs had declined to just 18.2 percent of the total cost of car ownership. (In 2004, gasoline costs were just 11.6 percent of total car ownership costs, an all-time low for the modern era.[6]) In fact, fuel costs are almost incidental. The reason is

FIGURE 20 Inflation-Adjusted U.S. Gasoline Prices, 1919–2007

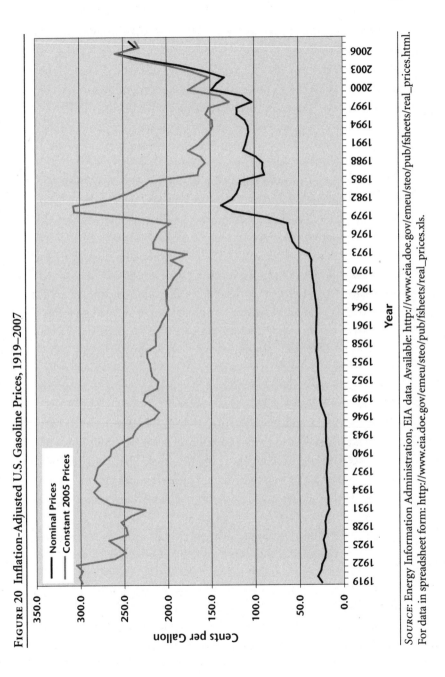

SOURCE: Energy Information Administration, EIA data. Available: http://www.eia.doe.gov/emeu/steo/pub/fsheets/real_prices.html. For data in spreadsheet form: http://www.eia.doe.gov/emeu/steo/pub/fsheets/real_prices.xls.

that the fixed costs of car ownership—insurance, licensing, taxes, and financing costs—have increased nearly fivefold since 1975.[7] Given the relatively low price of fuel, it's not surprising that Americans are opting for big vehicles with powerful engines. Indeed, in the overall cost of owning a car in the U.S., fuel expenses are not that significant.

While Americans are accustomed to inexpensive motor fuel, they are paying more for their fuel than necessary because of all the localized mandates on motor fuel formulations.

America has the most balkanized motor fuel market on earth. Refiners are now producing about 45 different blends of gasoline and multiple blends of diesel fuel.[8] And thanks to the new ethanol mandates, the mix of fuels—and the infrastructure needed to support them—has grown even more complicated. All of these regional blends of motor fuel increase the cost of production and transportation for motor fuel. In 2005, the Government Accountability Office found that there are 11 special blends of gasoline, which "are often used in isolated pockets in metropolitan areas, while surrounding areas use conventional gasoline." This proliferation of different types of gasoline is largely due to state-based efforts to address air quality issues. And while that may be a laudable goal, the result, according to the Government Accountability Office, has been higher costs for consumers. It studied gasoline prices in 100 cities and found that "the highest prices tended to be found in cities that use a special gasoline blend that is not widely available in the region, or that is significantly more costly to make than other blends." And consumers are paying those higher prices even though the agency determined that the air quality benefits of the myriad gasoline blends are "uncertain."[9]

Refiners have been agitating for more uniform fuel regulations for years. Their rationale makes sense: Simplifying the mix of fuels sold among the 50 states would help lower the overall cost of production and distribution and result in lower prices for consumers. It would also provide more energy security in the event of a supply disruption, as motor fuel from multiple refineries could be shipped to the region suffering shortages. Under the current system, supply disruptions are

What about Fuel Taxes and Carbon Taxes?

Over the last few years, I have written several articles that argue in favor of higher motor fuel taxes. I've also supported the idea of a carbon tax, which would impose levies on fuels based on their carbon content; that is, coal would be taxed more heavily than oil or natural gas.

Today, I doubt that either motor fuel taxes or carbon taxes are viable. In fact, they may be worse than doing nothing at all.

On the surface, higher fuel taxes appear to make sense. Higher prices should lead auto buyers to choose more fuel-efficient vehicles. The two biggest U.S. automakers have voiced support for fuel taxes. And over the past couple of decades, it's become obvious that federal mandates on fuel efficiency in the automotive fleet haven't been effective.

Higher taxes would send a clear market signal to motorists that their fuel costs are unlikely to be going down. It would remind them that if they want a larger vehicle, they can have it, but it will cost them. That type of message is preferable to command-and-control rules like the Corporate Average Fuel Economy standard, which has already been shown to be ineffective against the hordes of lobbyists fielded by the automakers. Indeed, ever since the mileage standards were created by Congress, the automakers have used various techniques to circumvent the rules and thereby build ever larger numbers of SUVs and trucks. The best example is the FFV ruse, under which GM, Ford, and Chrysler are making lots of noise about their love of corn ethanol and "renewable" fuels. The end result, though, has been higher oil consumption, as the FFV classification has allowed the automakers to artificially inflate their mileage numbers. The automakers used a similar dodge by classifying their minivans as "light trucks"—a ruse that, again, allowed them to make their fleets look more efficient than they actually were.

Higher fuel taxes would still allow consumers to buy the vehicles that they like, while freeing the automakers from more regulation. And given the struggles of the American carmakers, the last thing they need is more government mandates. The two biggest American carmakers have already expressed their preference for fuel taxes over new mandates on fuel efficiency. For instance, Ford Motor Company has supported a $0.50 increase in gasoline taxes. In 2004, Ford's chairman, William Clay Ford, Jr., told the *New York Times* that anything that "can align the individual customer's purchase decisions with society's goals are the way to go." GM's chairman and CEO, Richard Wagoner, concurred, saying, "If you want people to consume something less, the simplest thing to do is price it more dearly."[10]

continues

What about Fuel Taxes and Carbon Taxes *continued*

American motorists pay far less for their motor fuel than their counterparts in the developed world, and U.S. motor fuel taxes (which account for about 15 percent of the total cost) are the lowest in any industrial nation.[11] In early 2006, according to the International Energy Agency, U.S. drivers were paying $0.62 per liter of gasoline, while drivers in the UK were paying $1.55, Turkish drivers were paying $1.96, German drivers were paying $1.51, and Dutch drivers were paying $1.66.[12]

So how high would fuel taxes have to be to make a significant impact on overall oil use? Estimates vary, but several sources in the oil industry believe that gasoline prices would have to be at least $4.50 per gallon—or more—to effect a major change in American driving habits. Given mid-2007 prices of $3 or so per gallon, that would mean federal and state gasoline taxes—which now average about $0.47 per gallon—would have to triple, to $1.50 or more per gallon, in order to effect a significant reduction in overall demand. (The federal fuel tax rate is currently $0.184 per gallon. It has been at that level since 1993.[13])

But fuel taxes could cause as many problems as they cure.

Among the biggest problems is that higher taxes are regressive. America's poor would get hit much harder by new energy taxes than the rich. Lower-income consumers would pay far more as a percentage of their income than would the wealthy. Fuel taxes would also take a big toll on rural residents, who often must drive long distances for medical care, grocery shopping, and other essentials.

Another problem: Demand in the U.S. is fairly inelastic. America's sprawling cities—combined with the lack of good public transportation—leave many Americans no choice but to drive to their various destinations. Thus, higher fuel taxes would merely be a punitive levy that would do little to change behavior. Historical data show that despite significant price increases, gasoline demand has continued upward. For instance, in 1998, regular gasoline sold for about $1.03 per gallon.[14] That year, Americans consumed an average of 352.5 million gallons (8.4 million barrels) of gasoline per day.[15] By May 2007, gasoline prices were consistently above $3 per gallon—triple the 1998 price—and yet consumption that month exceeded 385.6 million gallons (9.18 million barrels) per day, an increase of almost 9.3 percent over 1998 levels.[16]

Another problem: How would the government return all of that new tax revenue to the people? An increased fuel tax would raise enormous amounts of new tax money. For instance, a $2-per-gallon gasoline tax would yield some $280 billion per year. Would Congress then allocate that tax to other issues, like, say, infrastructure, or would it seek to make the tax revenue neutral? If it were made neutral, would the government lower the income tax rates?

Another problem: Americans hate the idea of fuel taxes. In April 2007, a poll done by CBS News and the *New York Times* found that there was overwhelming support—92 percent!—among voters for requiring automakers to produce more energy-efficient cars. But when asked if they would support a tax on gasoline in order to "cut down on energy consumption and reduce global warming," 58 percent of the respondents said they were opposed. And 76 percent said they opposed a $2-per-gallon tax on gasoline.[17]

There are similar problems with a carbon tax.

A carbon tax (which would necessarily include a levy on motor fuel) would hit the poor much harder than the wealthy. It would raise the price of nearly every good and service. Electricity prices would rise because more than half of America's electric power is produced from coal. And while many economists believe that a carbon tax is the most logical way to address the issue of rising carbon dioxide emissions, the key problem is that carbon taxes need to be imposed internationally so that individual countries are not disadvantaged from a cost-of-energy standpoint. Without an international consensus, the tax regime will almost certainly fail to achieve its objectives.

This point was made clear in a July 2007 report by William Nordhaus, an economics professor at Yale University. The report, "The Challenge of Global Warming: Economic Models and Environmental Policy," advocates the imposition of a carbon tax that would start at about $30 per ton (that translates to a tax of about $0.30 per gallon of gasoline) and would be raised by 2 or 3 percent per year.[18]

Nordhaus writes that his computer models "point to the importance of near-universal participation in programs to reduce greenhouse gases."[19] If only half the countries of the world participate in the carbon tax program, Nordhaus estimates that there will be an "abatement cost penalty of 250 percent." In other words, unless everybody joins the carbon tax program, the countries that have agreed to tax energy will need to more than double their carbon tax rates in order to compensate for the countries that are not taxing energy.

Nordhaus goes on, saying that for his plan to work, "all countries would agree to penalize carbon emissions in all sectors at an internationally harmonized carbon price or carbon tax."[20] (It's worth noting that Nordhaus estimates that imposing carbon taxes high enough to

continues

achieve a 90 percent reduction in carbon dioxide emissions by 2050—the level advocated by Al Gore—would "impose a tax bill of $1,200 billion on the U.S. economy."[21])

So to summarize: Carbon taxes might help reduce carbon dioxide emissions, but to achieve those reductions, all of the world's countries would have to agree to (1) participate in such a program and (2) a price.

But is it realistic to expect agreement on those two issues? The answer appears to be a resounding no. By August 2007, 38 countries had still refused to join the international treaty to ban land mines.[22] If getting international consensus on a noneconomic issue like land mines is difficult, then imagine how hard it will be to forge an agreement to tax energy, the commodity that drives the world economy.

There is also a moral issue to the carbon tax question: Given the need for an international consensus to raise energy taxes, is it moral to impose such a levy on the world's poor? Today, 1.6 billion people do not have access to electricity in their homes. Some 2.5 billion people— about 40 percent of the world's population—use wood, dung, or other biomass to meet their cooking-energy needs. According to the World Health Organization, about 1.3 million people per year, most of them women and children, die because of the pollution caused by indoor biomass stoves. Only HIV/AIDS, malnutrition, lack of clean drinking water, and poor sanitation are greater health threats than the problems of polluted indoor air. As Fatih Birol, the chief economist for the International Energy Agency, has declared, "Decisive policy action is needed urgently to accelerate energy development in poor countries as part of the broader process of human development. We can not simply sit back and wait for the world's poorest regions to become sufficiently rich to afford modern energy services. . . . Access to energy is a prerequisite to human development."[23]

Given the pressing need for more fossil fuel energy among the world's poor, it will be morally perilous for the U.S., or any other country, to advocate a carbon tax—no matter how small—on countries with large numbers of people living in energy poverty.

Whatever gets decided with regard to tax policy, it's clear that there are no simple or cheap ways to reduce the amount of oil used in the U.S. Polls may show that voters like the idea of energy independence, but most American drivers appear perfectly happy with the status quo. And the likely result is a continuing embrace of big vehicles, low fuel taxes, and increasing oil imports.

difficult to overcome, particularly in metropolitan regions that require boutique gasoline blends.

REJECT THE CULTURE OF FEAR AND
ENGAGE THE ARAB AND ISLAMIC WORLDS

Fear sells. And in order to keep selling the long war, George W. Bush and his neoconservative cronies must continually remind their subjects in the U.S. and elsewhere that they are in dire straits, that there's an unshaven, dishdash-wearing jihadist from Yemen or Saudi Arabia lurking in every airplane lavatory, and that Islamic fanatics are prowling every airport, ready to cook up a Gatorade bomb that will kill dozens of innocents at the drop of a piece of carry-on luggage or pocket comb.

The continuing use of fear as a political tool—along with the constant drumbeat of terrorism—has become part and parcel of America's demented approach to energy policy.

It's abundantly clear that terrorism, when viewed dispassionately, has become a tool that politicians use to instill fear in the populace. That point was made in 2004, in a study by Ohio State University professor John Mueller that was published by the Cato Institute: "For all the attention it evokes, terrorism actually causes rather little damage and the likelihood that any individual will become a victim in most places is microscopic." Mueller then quoted Frantz Fanon, a 20th-century revolutionary, who said that "the aim of terrorism is to terrify." Given that truth, Mueller declared that "terrorists can be defeated simply by not becoming terrified—that is, anything that enhances fear effectively gives in to them."

Mueller argued that while many people worry about being killed in a commercial airplane crash, the risk of dying in that type of crash is 1 in 13 million. And that calculation takes into account the deaths that occurred on September 11, 2001. Mueller did not trivialize the risk of potential attacks from chemical or biological or nuclear weapons, but he said those weapons are "notoriously difficult to create, control and focus (and even more so for nuclear weapons)." He went on to say that while poison gas was a major worry during World War I, it accounted for less than 1 percent of all combat deaths and that it took an average of 1 ton of poison gas to produce one fatality.

Mueller skewered the media for their preoccupation with terrorism and the relative risks that it poses. He said the media have a "congenital incapacity for dealing with issues of risk and comparative probabilities—except, of course, in the sports and financial sections."[24]

While the media hype every study of bin Laden and every arrest of suspected terrorists, the statistics continue to show that terrorism poses little actual risk to most people. For proof, look at the U.S. State Department's own research. In April 2006, the State Department reported that in all of "2005, 56 Americans lost their lives in acts of international terrorism, and of that number, 47 of the fatalities occurred in Iraq."[25] In April 2007, the State Department reported that in 2006, only 28 U.S. citizens were killed worldwide in acts of terrorism. And of those 28 deaths, 22 occurred in Iraq.[26] Thus, only 6 Americans were killed by terrorists in locations other than Iraq.[27]

So let's compare the danger of terrorism with another risk: death by lightning strike. According to the National Weather Service, about 62 people are killed by lightning every year in the United States.[28] In 2006, there were 47 confirmed deaths from lightning strikes.[29] Thus, in 2006, about eight times as many Americans were killed by lightning as were killed by acts of terrorism outside Iraq.

Of course, it's foolhardy to ignore the risks of terrorism. And it would be callous and downright idiotic to say that the tragedies of September 11, 2001, should be ignored. It would be equally numbskulled to suggest that there won't be another attack, or that al-Qaeda will not be able to obtain a dirty nuclear bomb, or chemical weapon, or some other weapon of mass calamity.

That's not what's being argued here.

The U.S. must put terrorism into context and accept that it will never be eradicated. Disgruntled groups will always use violence against civilians for political purposes. Terrorism is a technique of warfare, not an ideology. Someone, somewhere, is always going to be willing to use violence for a political purpose, and the cheapest way to do that is through terrorism.

Let's further agree that fighting Islamic extremists requires more personnel who are focused on human skills like language and cultural studies. The problem is that the U.S. faces a dire shortage of those personnel. And that shortage exacerbates the risk that the U.S. faces from terrorism. Instead of taking a human-skills-based approach to fighting Islamic extremists, the U.S. military and the intelligence services are spending billions of dollars on super-high-tech airplanes, satellites, and other gizmos that provide no cultural context or human intelligence. In fact, the entire U.S. government faces a dire shortage of qualified Arabic speakers. That point was made clear in the December 2006 report by the Iraq Study Group, which found that the U.S. embassy in Baghdad—which has some 1,000 State Department personnel (that number does not include all of the service and security workers)—has just 33 people who can speak Arabic at all, only 6 of whom are fluent.[30]

The need for cultural knowledge is critically important given America's strategy for victory in the long war. That strategy is delineated in the U.S. military's 2006 Quadrennial Defense Review, a document that uses the word *terrorism* 38 times and *terror* 7 times. The QDR declares that "victory will come when the enemy's extremist ideologies are discredited in the eyes of their host populations and tacit supporters, becoming unfashionable, and following other discredited creeds, such as Communism and Nazism, into oblivion. This requires the creation of a global environment inhospitable to terrorism." The report continues, saying that achieving this goal will require "legitimate governments with the capacity to police themselves" and "effective representative civil societies."[31]

But how can the U.S. expect to build and foster those legitimate governments and civil societies if it lacks the personnel—and the ability to deploy them to the countries that need them the most? Without linguists, academics, diplomats, engineers, judges, lawyers, bankers, and other citizens who can help build civil societies by creating courts, building power plants, starting schools, lending money, and doing the hard work of nation building, the U.S. cannot achieve victory in the long war.

That very point was made by a Pakistani general, Bashir Baz, who was discussing the work done by Greg Mortenson and the Central Asia Institute described in the 2006 book *Three Cups of Tea*, written by Mortenson and David Oliver Relin.

Since 1996, Mortenson, a former mountaineer, has built more than 50 schools in Pakistan and Afghanistan. The schools have helped educate more than 22,000 children—many of them girls. In fact, Mortenson insists that all of the schools he works with educate girls. The reason: Education is a critical element in raising living standards in developing countries, particularly in the Islamic world. In spite of constant budgetary constraints (the Central Asia Institute's budget is about $1 million per year), Mortenson has had sparkling success mainly because of his grit, guts, language skills, and huge heart. And he's built all these schools without using any money from the U.S. government.[32] In *Three Cups of Tea*, General Bashir talks about Osama bin Laden, terrorism, and the conflict between the West and Islam. He said that America's enemy is "not Osama or Saddam or anyone else. The enemy is ignorance. The only way to defeat it is to build relationships with these people, to draw them into the modern world with education and business. Otherwise the fight will go on forever."[33]

Bashir's point is critical: The way to deal with the Islamic world is through engagement, by drawing them into the modern world and increasing the amount of educational, business, and cultural opportunities that are available to them.

In short, the real risk to America's security and prosperity isn't terrorism. Instead, the danger comes, largely, from the war on terrorism. And by conflating the war on terrorism with America's ongoing need for energy, the neocons and their supporters are further weakening America's hand when it comes to engaging the oil-rich countries of the Middle East. Those countries must be engaged in and brought into the world with "education and business."

The ongoing development and modernization of the Islamic world should be viewed as a positive development. Increasing wealth and better standards of living are good for everyone.

REDEFINE ENERGY SECURITY

Few phrases in the energy lexicon are as abused as "energy security." The phrase gets tossed around all the time, but few people agree on what it means. The best definition comes from A. F. Alhajji who, in late 2007 defined it thusly: "Energy security is the steady availability of energy supplies that ensures economic growth in both consuming and producing countries with the lowest social cost and lowest price volatility."

Alhajji's definition echoes the conclusions reached by the National Petroleum Council in its July 2007 report on future energy supplies. The report said that "energy independence is not realistic in the foreseeable future, whereas U.S. energy security can be enhanced by moderating demand, expanding and diversifying domestic energy supplies, and strengthening global energy trade and investment. *There can be no U.S. energy security without global energy security*" (italics added).[34]

In other words, America's ability to secure the energy it needs hinges on a stable, prosperous, global marketplace and global economy in which all of the players—producers and consumers alike—are able to get the energy they need at reasonable prices. Energy security means accepting energy interdependence. It means moving past the racism and xenophobia that dominate the current energy rhetoric and embracing the global market. It means helping the developing countries of the world build energy infrastructure so they can develop their economies and raise the living standards of their people. Unless and until those things happen, the U.S. won't have energy security.

ACCEPT INCREASING ENERGY USE AND ADAPT TO A CHANGING GLOBAL CLIMATE

In 2006, China's electric grid added France.

Or, to be more precise, in 2006, China expanded its electric generation capacity by 102 gigawatts—that's about the same capacity as all of France's electric power plants *combined*.[35] What makes China's 102 gigawatts of new power capacity so extraordinary is that nearly 90 percent came from coal-fired power plants.[36] China's electric power sector is going to continue growing, and most of that growth will come from coal-fired power plants. By the end of 2006, China had about 622 gigawatts of electric generation capacity.[37] For comparison, the U.S. had about 978 gigawatts of capacity and about one-fourth as many people as China.[38] But China is racing to catch up. By mid-2007, it was adding a new coal-fired power plant every four days.[39]

The fact that China is adding such enormous quantities of new electricity capacity underscores the fact that the world's developing countries need ever-increasing amounts of energy to develop their economies and improve the living standards of their citizens. And as those countries continue to grow, their energy use—and therefore, their carbon dioxide emissions—is going to continue increasing.

Again, this book is not arguing one side or the other in the global warming debate. That said, it's obvious that a resident of Los Angeles who drives a hybrid car will do little to reduce overall global carbon dioxide emissions as long as his or her counterpart in Shanghai is increasing his or her consumption of coal, or oil, or natural gas. And that is exactly what's happening. Given these realities, and the fact that energy use continues ever upward in nearly every country on the planet, we likely have no choice but to adapt to the changing global climate for the simple reason that curbing carbon dioxide emissions to any significant degree appears hopeless.

There are plenty of data to support that point.

Between 2001 and 2005, China's coal consumption doubled to some 2.1 billion tons.[40] By 2006, China was burning more coal than the U.S., the E.U., and Japan *combined*. Pollution from those plants was showing up as far away as the mountains of California and Oregon. And even though coal is causing acid rain in China and contributing to environmental problems in other countries, China's going to continue burning ever-increasing amounts of coal in order to fuel its economic growth.[41] In 2006, that coal consumption allowed China to surpass the U.S. in total carbon dioxide releases. According to the Netherlands Environmental Assessment Agency, China emitted 6.2 billion tons of carbon dioxide in 2006, eclipsing the 5.8 billion tons released by the U.S. that year.[42]

FIGURE 21 The Most Populous Countries Ranked by Per Capita
Electricity Consumption

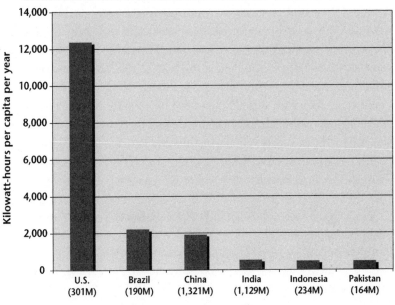

Note: Population numbers in parentheses, in millions.

SOURCE: Nationmaster.com. Two data sets were used for this chart. Available:
http://www.nationmaster.com/graph/peo_pop-people-population, and http://
www.nationmaster.com/graph/ene_ele_con_percap-energy-electricity
-consumption-per-capita.

India, the second-most-populous country on the planet, is emulating
China's electrification plans. By 2030, India plans to more than triple its
electricity generation capacity, going from about 130 gigawatts in 2007
to about 400 gigawatts.[43] And as in China, the vast majority of that new
electricity will be generated by burning coal. By 2012, India plans to add
more than 46 gigawatts of new coal-fired power plants.[44] That 46
gigawatts is approximately equal to all of the electricity-generating capacity in Mexico.[45] India's electricity demand is driving its coal consumption. By 2012, India's coal consumption is expected to jump by
more than 50 percent to some 730 million tons per year.[46]

But even with the addition of huge numbers of new power plants, India and China have a long way to go before their citizens will be able to enjoy the quantities of electricity that are used by Americans.

In addition to demand for electricity, growing prosperity around the world has spurred the use of air travel. Between May 2006 and May 2007, the global airline industry grew by 5 percent to some 2.5 million commercial airline flights per month. In China, scheduled commercial flights for May 2007 jumped by 18 percent, compared to flights in May 2006. During that same period, flight numbers in India jumped by 25 percent, and in Spain, they increased by 16 percent.

The growth of air travel in the developing world is almost certain to continue because most countries have far fewer commercial aircraft than the U.S. And as the air travel in these developing countries grows, so, too, will the demand for jet fuel. The disparity in available aircraft can be seen in Figure 22.

This surge in air travel is part of the global economic boom. But some scientists believe that carbon dioxide emissions from aircraft, which are released at altitude, are 2.5 times more damaging in terms of potential global warming than emissions that come from cars or electric power plants. Further, even though the aviation sector is among the fastest-growing sources of greenhouse gases, there is no practical way to cut those emissions because large segments of the aviation business are considered international activity and are therefore excluded from the Kyoto Protocol.[47]

Meanwhile, carbon dioxide releases in the U.S., Europe, and Japan continue to increase.

Between 2000 and 2005, U.S. carbon dioxide emissions rose by 2.5 percent, and European emissions jumped by 3.8 percent.[48] And although many European leaders have bashed the U.S. for its reluctance to sign the Kyoto Protocol, only 5 of the 15 member countries of the European Union that initially agreed to the Kyoto measures are on track to meet their targets for carbon dioxide reductions.[49] (The E.U. has since expanded twice, in 2004 and again in 2007. It now has 27 members.)

Pointing out that the E.U. is not meeting its carbon reduction targets is meant not to demean the E.U., but to underscore the fact that

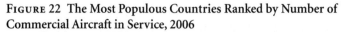

Figure 22 The Most Populous Countries Ranked by Number of Commercial Aircraft in Service, 2006

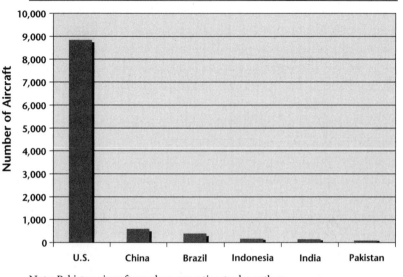

Note: Pakistan aircraft numbers are estimates by author.

SOURCES: Frost and Sullivan, "North American Commercial and Military Ground-Based Flight Simulation Market, August 31, 2006, 5-1 to 5-3.

obtaining the drastic cuts in carbon dioxide emissions envisioned by politicians like Al Gore—who wants a 90 percent cut by 2050—verges on the impossible, without massive economic costs. In fact, achieving the far more modest cuts outlined by Kyoto also appears to be beyond the ability of most countries. Spain provides a good example. In 2004, according to the European Environment Agency, it was exceeding its Kyoto reduction targets by a whopping 31 percentage points. Denmark, which has been lionized around the world for its embrace of wind power, was exceeding its Kyoto targets by nearly 8 percent.[50]

Japan, the home of the Kyoto Protocol, is also not meeting its reduction targets. In 2004, it was exceeding its Kyoto targets by 19 percent. If Japan, arguably the world's most homogeneous and resource-constrained industrial society—as well as one of its most environmentally focused

countries—has not been able to meet its Kyoto goals, it seems highly unlikely that the U.S. will be able to make any significant reductions in its
carbon dioxide output. The reality of Japan's situation was laid out in
February 2007 by Kiichiro Ogawa, a research fellow at Japan's Institute of
Applied Energy, when he wrote in the *Energy Tribune* that renewable energy could not become a major source of energy in Japan, and that "fossil
energy, that is to say, oil, coal, and natural gas, will be the major energy
sources for the time being, and probably through 2050."[51]

If—*and that's a very big if*—the countries of the world are truly interested in reducing carbon dioxide emissions, they will have to work
together. If they don't agree to some kind of global plan, then all of the
talk about global warming and greenhouse gases just adds more hot
air to the troposphere.

The hard facts show that carbon dioxide emissions continue climbing because energy consumption—and particularly fossil fuel consumption—goes hand in hand with economic development. Given
that reality, the residents of the planet will likely have no choice but to
adapt to a changing global climate—*whatever the cause*—and like it or
not, that adaptation process will require the use of large amounts of
fossil fuels. And while fossil fuels will continue to dominate the
world's fuel mix, the U.S. should also look at more low-carbon
sources. And that leads to the next idea.

EMBRACE SOLAR AND NUCLEAR, AND
PURSUE NEW TECHNOLOGIES AND EFFICIENCY

If the countries of the world agree that carbon dioxide emissions are a
problem, then more investments should be made in the most promising sources of low- or no-carbon electricity: solar and nuclear power.

Between 2001 and 2004, the amount of installed solar power around
the world more than doubled. In 2005 alone, global production of solar
cells jumped by 45 percent.[52] Although solar power faces many of the
same limitations as wind power (the obvious one being the intermittent nature of sunlight, cloudy days, etc.), it also has many advantages.

Photovoltaic panels can be placed on existing buildings and thus don't compete for available land, as is the case for biofuels and wind turbines. Solar panels can cut electricity demand during the peak usage periods, such as on hot summer days when air-conditioning demand is highest. Photovoltaic panels have no moving parts, create no noise, and can generate electric power for decades with no emissions. (Obviously, there are other solar applications that go beyond photovoltaics. These range from passive solar designs in homes to large-scale thermal electric power plants, which use mirrors to concentrate the sun's rays. While those are useful technologies, the photovoltaic technology appears to have the most potential for widespread deployment.)

As discussed earlier, solar power remains far too expensive to compete with fossil fuels on a head-to-head basis. But significant progress is being made. One promising new technology turns infrared light into electricity, thereby dramatically improving the efficiency of existing photovoltaic cells. About half the energy in sunlight is found in the infrared segment of the light spectrum. By capturing both the visible and the infrared parts of the spectrum, photovoltaic panels could capture up to 30 percent of the sun's radiant energy.[53] In late 2006, Spectrolab, Inc., a subsidiary of defense giant Boeing, announced that it had demonstrated a photovoltaic cell that can convert 40 percent of the sun's energy into electricity.[54]

Despite the promise of solar, the American government is investing precious little in developing new sun-powered technologies. In 2007, the U.S. Department of Energy spent just $159 million on solar research and development.[55] That's less than half the amount that the same agency has given away in the form of grants to companies that are trying to develop cellulosic ethanol. But this research—and the spread of solar technology in energy-poor countries—should be actively funded by private foundations and not just government agencies.

While solar will likely provide only small amounts of new generation capacity, nuclear power can provide large increments of low- or no-carbon electricity to the world's energy mix over the next couple of decades. Nuclear is the only sector that has enough momentum and enough capital behind it to make a significant dent in the overall use of fossil fuels. And over the past few years, there have been important breakthroughs in nuclear technology. Toshiba has been developing a small nuclear plant that puts out just 10 megawatts of power. It has no moving parts and is cooled by liquid sodium instead of water. Using sodium allows the reactor to run hotter and avoid using highly pressurized pipes. Initial estimates found that the reactor could produce power for between $0.05 and $0.13 per kilowatt-hour, which would be far cheaper than generators that rely on large diesel engines.

The Toshiba design is just one of many nuclear technologies now being developed. Others include the pebble-bed modular reactor, which uses helium as a coolant. In addition, both General Electric and Westinghouse have submitted new reactor designs to the U.S. Nuclear Regulatory Commission for its approval.[56]

Some of the world's leading environmentalists have decided that nuclear power is the best option for the future. Patrick Moore, a founder of Greenpeace, has become one of the most ardent backers of the nuclear option. In 2006, Moore wrote an opinion piece for the *Washington Post* saying that nuclear energy may be the "energy source that can save our planet." Although Moore now does public relations work for the nuclear power industry, he pointed out that many other ardent environmentalists, including Stewart Brand, one of the founders of the *Whole Earth Catalog,* and James Lovelock, the British scientist who came up with the Gaia theory about the resilience of the planet Earth, are advocates of nuclear power. In mid-2006, Moore told me that when it comes to producing large increments of new, low-carbon electricity, there "aren't really any other choices. Fossil fuels are still a large segment of our consumption. Coal, nuclear, and hydro are the only choices. That said, by and large, even with nuclear being aggressively pursued, it'll still be hard to reduce fossil fuel because China and the U.S. have such huge reserves of coal."[57]

The big American environmental groups remain adamantly opposed to nuclear power. In April 2007, a spokesman for the Sierra Club's global warming campaign declared that "switching from coal to nukes is like giving up smoking and taking up crack."[58]

Environmental groups oppose nuclear power largely because of their concerns about radioactive waste. While it's true that used nuclear fuel continues to be dangerously radioactive for centuries, that waste can be effectively managed. The U.S. government has spent some $7 billion building a repository for nuclear waste at Yucca Mountain in Nevada.[59] But the waste site has been stalled because of political objections. A leading opponent: Harry Reid, the Democratic senator from Nevada who is now the Senate Majority Leader. Reid has flatly declared that Yucca Mountain "is never going to open" and that it is "not the answer to nuclear waste storage." He prefers that all of the radioactive waste produced by America's reactors be stored at the plants where it is produced.[60]

Regardless of how and where spent nuclear fuel is stored or reprocessed, the issue for nuclear power is the same as for any other energy source; that is, pick your poison. Environmental groups like the Sierra Club don't like carbon dioxide emissions. Nor do they like nuclear power. The problem is that every energy source has an impact. In the case of carbon dioxide, it goes up into the atmosphere, where it cannot be managed. Nuclear waste stays down here on the earth, where it can be managed and watched. And the quantity of waste involved is tiny. A 1,000-megawatt nuclear plant produces about 1 metric ton of waste per year. A coal-fired power plant of that same size produces about 1 million tons of waste.[61]

Given those quantities, and the ongoing push by environmental groups and the European Union to drastically cut carbon dioxide emissions, managing a relatively small volume of nuclear waste seems eminently doable, given, of course, the political will to do so.

Another promising technology: biodiesel produced from algae. Although the technology has not been demonstrated on a large scale, early results show that it can be used to sequester carbon dioxide while also producing plant material that can be used to make diesel fuel and ethanol. Researchers claim that algae ponds can theoretically produce 15,000 gallons of biodiesel per acre—about 250 times as much as the amount from a comparable acre of soybeans.[62] By one calculation, 15,000 square miles of algae ponds located in the southwestern U.S.

could produce enough algae to supply all of America's diesel fuel needs.[63] Early tests at a power plant owned by Arizona Public Service have been promising.[64] Of course, as with every new energy concept, many hurdles must be overcome. Demand for freshwater could limit production. Thus, all claims about algae diesel's potential productivity must be taken with a large grain of salt.

While nuclear, solar, and algae-based fuel may add new sources of energy, none of them will be enough on their own. There must also be continuing efforts to improve the efficiency of appliances, automobiles, and industrial processes.

What else can be done? Well, congestion pricing—charging motorists for driving into traffic-heavy urban areas during key times of day—may reduce consumption. Another idea: mileage-based insurance policies, which would charge motorists for the amount that they drive, instead of the current system, which assesses drivers the same amount regardless of how much they drive.

Congress could offer tax incentives to buyers of high-mileage cars. And while many people are promoting plug-in hybrid electric vehicles, Congress shouldn't limit incentives to hybrids. It could also extend them to buyers of diesel cars, which are about 30 percent more fuel-efficient than comparable gasoline-powered vehicles. Incentives could also be given to buyers of vehicles that use propane or compressed natural gas instead of gasoline.

All of these efforts—efficiency, solar power, nuclear plants, algae diesel, and every other tool in the shed—should be used to provide more energy to the world's economy. But there should also be ongoing efforts to find a breakthrough technology. And few technologies would be more revolutionary than a superbattery.

CREATE THE SUPERBATTERY PRIZE

For more than a century, batteries have confounded some of the world's greatest scientists. Thomas Edison spent more than a decade and invested some $1.5 million (about $32 million in present-day money) of his own money trying to perfect a battery that would make electric cars viable. He failed.[65] Today, batteries are even more important and have many more uses than in Edison's day.

It's easy to understand why Edison and others have worked so hard to improve the energy density of batteries. A super-high-capacity electricity storage system would revolutionize the energy sector. It would be, as one engineer told me, the "silver bullet." The applications of such a system are obvious:

- Renewable energy sources like solar and wind would be far more viable as the intermittency problem would largely be resolved.
- Base-load coal-fired power plants could charge batteries at night, when demand is low. During peak demand periods (like on hot summer afternoons), they could feed that stored electricity into the grid, thereby allowing far better fuel efficiency.
- Electric cars, long hampered by their limited range, would become far more viable.
- The electric grid would become "smarter" and more resilient. With thousands, or perhaps millions, of high-capacity super-batteries distributed across the country, individual businesses and even homeowners could buy and sell electricity to the local grid. Blackouts due to falling tree limbs or power plant outages would be minimized because locally stored electricity could be used to replace the missing power.

So how do we push for a revolutionary breakthrough in battery technology? The answer: the Superbattery Prize.[66]

Offer $1 billion to the inventor of a super-high-capacity system that is compact, affordable, and capable of storing multiple kilowatt-hours of electricity. Offer $10 billion to the inventor of a system with all of those attributes that can also store multiple megawatt-hours of electricity. The money and refereeing for the Superbattery Prize could be provided by either the U.S. Department of Energy or, better yet, a consortium of private foundations.

There is plenty of precedent for using prize money to drive innovation. In 1919, a hotelier named Raymond Orteig offered $25,000 to the first pilot who could fly nonstop between New York and Paris. Eight years later, a previously unknown American named Charles Lindbergh collected that prize.[67]

More recently, aviation whiz Burt Rutan, backed by billionaire Paul Allen, collected the $10-million Ansari X Prize, which was offered to the first privately built vehicle that could fly to the edge of space, return to earth, and repeat the feat within two weeks. In 2004, their creation, SpaceShipOne, made two trips into low-earth orbit, and Rutan and Allen claimed the $10-million prize. In doing so, they ushered in an era of privately financed spaceflight.[68]

In early 2007, British billionaire Richard Branson offered a prize (called the Virgin Earth Challenge) that relates directly to energy usage and carbon dioxide emissions: He offered $25 million to anyone who could invent a technology that would remove 1 billion tons of carbon dioxide per year from the atmosphere.[69] And while Branson's prize is laudable, a better solution would be one that prevents the release of carbon dioxide. That's why super-high-capacity batteries are so desirable: They would save enormous amounts of fossil fuels and thereby prevent the release of huge amounts of carbon dioxide.

The problem with batteries has always been energy density. As the graphic in Figure 23 shows, even the latest lithium-ion batteries cannot come close to matching the energy density of liquid fuels like ethanol, gasoline, and diesel fuel. Nor can they match the energy density of compressed hydrogen or compressed natural gas. Of course, as Edison and many others have learned, electrons don't like to slow down. And despite more than a century of work, today's batteries remain similar in design and energy density to the lead-acid and alkaline batteries that Edison worked with in his lab.

But the lure of $1 billion—or $10 billion—should motivate scientists and tinkerers to at least look at the Superbattery Prize. And if they succeeded in creating a superbattery, the world would be altered in ways even more profound than the changes that ensued after Lindbergh collected that $25,000 from Orteig.

But even high-capacity batteries won't solve the need for increasing supplies of energy, particularly the type needed for transportation.

FIGURE 23 Energy Density of Batteries versus Gaseous Fuels and Liquid Fuels (Gasoline = 10)

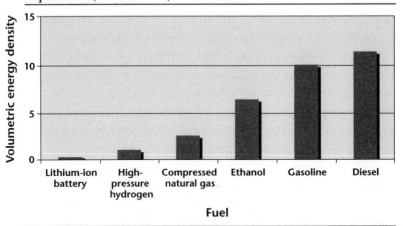

SOURCE: Toyota Motor; Takehisa Yaegashi, "Beyond Hybrids: Toyota's Approach towards the 'Ultimate Eco-Car,'" undated, presented at International Research Center for Energy and Economic Development, 2007, Boulder, Colorado.

INCREASE DOMESTIC OIL PRODUCTION

Few issues expose America's schizophrenia about energy issues better than the battle over the Arctic National Wildlife Refuge (ANWR). The refuge, which contains an estimated 10.4 billion barrels of oil, has become a live-or-die symbol for the major environmental groups in the U.S.

These environmental groups claim that the ANWR will supply only a small amount of America's oil needs and therefore should be left alone. But the U.S. Department of Interior estimates that the ANWR could produce about 1.4 million barrels of oil per day, which would make that field's output larger than that of Texas, America's foremost oil-producing state.[70] Given its huge potential and the desire to reduce oil imports, the refuge should be opened immediately to oil and gas exploration and production.

Alaska also has enormous supplies of natural gas that have yet to be tapped. Prudhoe Bay, the most famous oil field in Alaska, holds some 31 trillion cubic feet of natural gas.[71] That's enough gas to supply all of

America's natural gas needs for about 18 months.[72] And yet, wrangling over possible pipeline routes has prevented the development of Alaskan gas. Congress needs to move past the provincialism that has stalled the gas pipeline project and act in the best interests of all Americans.

Yes, the ANWR is an important piece of turf. And of course, oil development in the refuge will have consequences. But it's important to remember that primary oil development will occur on only a tiny portion (perhaps as little as 2,000 acres) of the refuge's 19.5 million acres.[73] Furthermore, it's worth remembering that the ANWR itself represents less than one-tenth of all the federal land in Alaska.[74]

And while the ANWR is an important bit of domestic territory for oil exploration, there's an opportunity that's likely just as big, or bigger, and it's closer to the continental U.S.: the offshore waters in the eastern Gulf of Mexico and the East Coast. For decades, federal lawmakers in coastal states and environmentalists have been happy to burn the oil and gas that comes from rigs located in offshore Texas and Louisiana. But when it comes to developing the proven oil and gas reserves located offshore in states like Virginia, Florida, or California, or the eastern Gulf of Mexico, the rhetoric quickly changes. Discussions about energy independence and the need to reduce oil imports are quickly discarded and are replaced by a more common stance: NIMBY."

This lack of consistency can be seen in the May 2002 deal in which the U.S. Department of Interior paid ConocoPhillips, Murphy Oil, and Chevron a total of $115 million *not to drill* for oil and gas on offshore leases they owned near Pensacola, Florida. Those leases contain some 700 billion cubic feet of natural gas. Perhaps it was just a coincidence, but the president's brother, Jeb, was running for reelection at the time of the deal.[75] That 700 billion cubic feet of gas—located in a formation known as the Destin Dome—would be enough to supply all of Florida's gas needs for about a year.[76]

Tussles between states and the federal government over offshore drilling rights aren't new. The Tidelands controversy, which raged throughout the late 1940s and 50s, pitted Texas against the federal government for control of 2.4 million acres of offshore territory claimed by the state. California was also involved. Before the deal was settled—in favor of Texas—there had been three U.S. Supreme Court decisions against the states, three acts of Congress in favor of the states, and two presidential vetoes against the states. Offshore drilling rights were a major issue in the 1952 presidential race between Dwight Eisenhower and Adlai Stevenson. Stevenson was against the states. The Texas Democratic Party was so incensed by Stevenson's opposition that it passed a resolution urging Democrats to vote against him in November. Eisenhower beat Stevenson in Texas and went on to win the White House.[77]

What's different today from the debates over the Tidelands is that the states now want control over what is unquestionably federal territory. The old fight between Texas and the federal government involved the territory that lay within three leagues (10.35 miles) of the low-tide line. (The federal government insisted the states owned only the territory within three miles of the shore.[78]) But by 2006, the anti-oil-development forces had become so strong that Congress was considering legislation that would give states the right to control the waters within 100 miles of the shore.[79] And some states were pushing for special exemptions. For instance, California's senators, Barbara Boxer and Dianne Feinstein, both Democrats, have pushed legislation that would permanently ban offshore drilling for oil and gas in their state. They have also asked for exemptions to a provision in the 2005 federal energy bill that requires all of the states to do an inventory of their oil and gas reserves.[80]

While states' rights are important, the oil and gas reserves in federal waters belong to the taxpayers of all 50 states, not just to those who live in the coastal states. By ceding power to the coastal states, citizens of the noncoastal states are being forced to forgo the royalty income that can be earned from those offshore assets. They are also forgoing the potential reduction in foreign oil imports.

How much oil is there? The Minerals Management Service estimates that the eastern Gulf of Mexico alone holds some 20.2 trillion cubic feet of gas and 3.6 billion barrels of oil. In addition to these huge resources, the eastern Gulf of Mexico is attractive because it's close to huge amounts of already-existing technology, personnel, and infrastructure. The western Gulf of Mexico—that is, the offshore areas adjacent to Texas and Louisiana—has been thoroughly developed, and it continues to see huge amounts of new drilling activity. More drilling in regions further east can be done quickly and therefore add substantial amounts of domestic production to America's energy portfolio. Furthermore, the eastern Gulf of Mexico is becoming more attractive for prospecting, particularly given the recent discoveries like Jack, the 2006 wildcat well in the deepwater off the shore of Louisiana, which found deposits that may ultimately contain 15 billion barrels of oil. It is plausible that the same productive geology that dominates offshore Louisiana extends further eastward, and that other huge oil and gas deposits may lurk in offshore Mississippi, Alabama, and Florida.

The prohibitions on drilling in locations like the ANWR and the eastern Gulf of Mexico are perhaps the most obvious examples of how the ongoing calls for energy independence ring hollow. Increasing the domestic production of fossil fuels will help reduce America's reliance on foreign sources. And yet, vast areas of federal territory are being kept off-limits under the guise of environmental protection.

Of course, some federally owned lands deserve to be protected. And they should be. But if the U.S. is truly serious about reducing its imports of foreign energy, it faces some hard choices. It cannot continue fencing off large tracts of hydrocarbon-rich territory while also insisting that foreign oil is bad. At some point, the U.S. is going to have to choose between rhetoric and practicality when it comes to domestic energy sources. But even as the rhetoric continues, the world is migrating toward cleaner fuels.

EMBRACE NATURAL GAS

Although many pundits talk about peak oil, there's not much talk about "peak gas." That's because the globe continues to have a surfeit of natural gas and more gas is being discovered all the time. Better still, the number of countries with known reserves increased from about 40 back in 1960 to more than 80 as of 2005.[81] Between 1980 and 2006, global gas reserves more than doubled. And along with the significant increases in consumption, new reserves continue to be found. Between 1996 and 2006, global proved natural-gas reserves increased by nearly 23 percent, to some 6,405 trillion cubic feet. At the current rates of extraction, that's enough gas to supply the world for another six decades.[82] Furthermore, new gas reserves are being found at a much faster rate than are new oil reserves.

The growing global supplies of gas—and the demand that's eager to make use of those supplies—are positive trends that should be encouraged for a number of reasons. First and foremost among them is that gas is the cleanest of the fossil fuels. It emits about half as much carbon dioxide as coal and creates far fewer air pollutants.[83]

Environmentalists should applaud the increased use of natural gas, as it is part of the ongoing "decarbonization" of the world's energy mix. This trend has been going on for about two centuries. And the trend can be understood from the ratio of carbon to hydrogen atoms in the most common fuels. From prehistory through, say, the 1700s and early 1800s, wood was the world's most common fuel. Wood has a carbon-to-hydrogen ratio (C:H) of 10 to 1. That is, wood has about 10 carbon atoms for every 1 hydrogen atom. But as the Western world industrialized during the 1800s and early 1900s, wood lost its dominance to coal. Coal was a dramatic improvement over wood because it is a far denser source of energy, with a C:H ratio of about 2 to 1. But coal was destined to lose out to oil as the fuel of choice, particularly for use in transportation, because of oil's superior energy density. Oil also has the virtue of being cleaner, a product of its C:H ratio of about 1 to 2. Over the coming decades, natural gas (CH_4) consumption will increase because it is the cleanest of the fossil fuels, thanks to its C:H ratio of 1 to 4. That is, it has just 1 carbon atom for each 4 atoms of hydrogen.

The inexorable decarbonization of the global economy is occurring because energy consumers are always seeking ever denser forms of energy to allow them to do ever greater and ever more precise amounts of work. (Lasers are a prime example of this trend toward superconcentrated energy forms.) Jesse Ausubel, the director of the program for the human environment at Rockefeller University in New York City, said the trend toward decarbonization may waver for a decade or two as countries like India and China add massive amounts of new coal-fired power plants, but "over the long term H gains in the mix at the expense of C, like cars replacing horses, colour TV substituting for black-and-white, or email gaining the market over hard copies sent through the post office."[84]

Second, while the decarbonization trend is important, it's just as important to realize that gas can be substituted for oil in many different applications. With some modifications, most standard automobiles can be converted to run on natural gas. Gas can be used to generate electricity and heat homes, and as a feedstock for an array of chemicals and other products.

Unfortunately, gas infrastructure is more expensive than oil infrastructure. Pipelines are inherently expensive and are constrained by geography. But the booming global market for liquefied natural gas, carried aboard massive ships, as well as the advent of a less expensive technology, compressed natural gas, could make a big difference. Compressing gas is far cheaper than liquefying it, which requires expensive cryogenic plants that can cost $1 billion or more. Although the compressed-natural-gas technology doesn't compress the gas as much as the cryogenic process, it is still effective and could thus be a significant player in the global gas market.

That should be good news for the U.S., which, despite huge investments in new drilling, continues to experience declining gas production. In May 2007, Robert Gillon, an analyst at John S. Herold Inc., issued a report on domestic gas production in which he pointed out that in the 1990s, the U.S. was drilling about 10,000 new gas wells per year and gas production was rising. "Now," wrote Gillon, "30,000 new [well] completions a year is inadequate."[85] Gillon made it clear that even with huge new investments in drilling, U.S. gas production will continue its long downward trend. That means the U.S. will have to rely on more imported natural gas. But that should not dissuade the U.S. from using more of this clean and abundant fuel.

Wood dominated the global energy scene through the 18th century. Coal dominated the 19th century. Oil dominated the 20th century. Given the ongoing decarbonization of the world's energy mix, natural gas should be the dominant fuel of the 21st century. And that dominance should be welcomed.

22

CONCLUSION

Energy is the most important commodity in the global economy. It is, as the economist Julian Simon once declared, the "master resource."

Given energy's importance, it stands to reason that discussions about energy would be rational, based on the best scientific analysis, and on careful assessments of available resources, their costs, and their potential benefits. But as the current craze about energy independence illustrates, rationality plays a secondary role in most energy discourse. As the Cato Institute's Jerry Taylor and Peter Van Doren wrote in May 2007, "Once the topic of conversation turns to oil, the human brain spontaneously short-circuits, nodal synapses fire randomly, and I.Q. points bleed out of the cranium and all over the kitchen floor."[1]

America needs to stop the bleeding and retrieve those IQ points and do so right damn quick. America needs to educate itself about the world's biggest enterprise. Every year, the world's residents spend about $5 trillion finding, refining, delivering, and consuming energy.[2] And yet, the extent of most Americans' energy knowledge goes no further than what can be contained by a snappy sound bite that bashes the Saudis or complains that gasoline now costs almost as much as milk. Simply put, energy is too important to the U.S. economy and the world economy for discussions about it to devolve into meaningless rhetoric. The U.S. needs to move beyond its current infatuation with government by cliché and face the reality that we live in an interconnected, interdependent world.

When it comes to energy, America stands at a crossroads. In one direction is the empty rhetoric of energy independence and the concomitant evils of protectionism and isolationism. In the other direction lies the modern world of engagement, politics, free markets, and the active embrace of globalism and all the complications that come with it.

By now, it should be obvious where America should focus its efforts. On that same road toward energy independence lurks the graveyard of military hegemony, the idea that the U.S. can dominate the world and do so based solely on its military might. The concept of energy independence is closely tied to the neoconservatives' belief that any type of engagement and interchange with the Arab and Islamic worlds must be avoided at all costs. The unfortunate truth about the drive for energy independence is that it is, at its most fundamental level, a xenophobic and—dare one say it?—*racist* response to a complicated world that has grown beyond America's ability to control.

The U.S. no longer controls the global market in oil, natural gas, uranium, or refined products. Those days are long gone. And yet, the calls for energy independence continue, as though merely wishing for a bygone era will make America strong again.

Despite the piffle coming from the neoconservatives, the Republicans, the Democrats, the pundits at the *New York Times,* and the environmentalists, the U.S. cannot be energy independent. Nor can it wage a never-ending war on the Islamic oil producers. Like it or not, the U.S. has to coexist with a bunch of unsavory oil-rich nations and the scoundrels who run them. But America can do so while still promoting freedom and democracy. The U.S. can buy the oil it needs and still despise the practices of the oil producers. And in all too many cases, the human rights practices of the world's biggest oil producers are abominable. That's been true for a long time. Oil rarely brings wealth to countries. Instead, it usually brings only poverty and war. That's clearly true in places like Iraq and Nigeria.

But that does not mean the U.S. can disengage from the rest of the world. Instead, America must overhaul its foreign policy, particularly with regard to the Islamic world. It must—as military analyst, author, and Boston University professor Andrew Bacevich wrote in August 2006—"be smart, and getting smart means ending our infatuation with war and rediscovering the possibilities of politics."[3]

Embracing the possibilities of politics also means embracing America's long-held beliefs in free markets and free trade. Instead of retreating into the protectionism and isolationism that come with the arguments for energy independence, the U.S. should embrace the global energy trade and accept that every nation should sell the things that it produces better (or more cheaply) than its competitors. That stance should result in greater economic interdependence, and that interdependence should reduce the risk of war.

Instead of wasting billions of taxpayer dollars on false fuels like corn ethanol, the U.S. should be investing in projects that will keep America vibrant. Health care costs are soaring. Public education is faltering. Entitlement spending is out of control. America's once-dominant technological lead in everything from computers to automobiles has largely evaporated. Transportation infrastructure is crumbling. In May 2007, the Urban Land Institute and the accounting firm Ernst & Young LLP released a report saying that there is an "emerging crisis" in infrastructure and that the U.S. faces a $1.6-trillion deficit in infrastructure spending through 2010. The U.S. now spends less than 1 percent of its GDP on infrastructure. By comparison, India spends 3.5 percent and China spends 9 percent.[4] While the statistics are eye-popping, the parlous condition of America's infrastructure was made all too real on August 1, 2007, when an interstate bridge over the Mississippi River collapsed in Minneapolis, Minnesota, killing 13 people.

Instead of continuing to waste billions of dollars and thousands of lives by continuing the Second Iraq War, the U.S. needs to focus more of its economic priorities on rebuilding its bridges, roads, and other transportation infrastructure. It must update its aging oil and gas pipelines and overloaded, antiquated electricity grid.

America must begin paying off its soaring debt. By September 2007, the U.S. government owed about $9 trillion.[5] In 2006 alone, the interest payments on that debt totaled a staggering $405.8 billion.[6] For comparison, the U.S. Department of Transportation's 2006 budget was about $60 billion.[7]

Instead of continuing the Second Iraq War, the U.S. must accept the notion that having access to the Persian Gulf's oil does not require total militarization of that region. The People's Republic of China is buying all of the oil it needs from the Persian Gulf countries, and it doesn't have a single soldier on the ground in Iraq, Kuwait, Saudi Arabia, or Iran. The U.S. has soldiers in three of those four.[8] By not militarizing the Gulf, the Chinese are saving billions of dollars. And they are gaining stature throughout the world, despite their lousy record on human rights and property rights. The Chinese, by not fighting elective wars, are growing stronger and richer. Meanwhile, America, by fighting an elective war, has dug itself into a quagmire that has made it weaker and poorer. And the longer the long war continues, the weaker and poorer America will become. Yes, the U.S. has long-term energy interests in the Persian Gulf. But so does every other country on the planet.

To put it simply, the U.S. cannot afford to continue deploying hordes of "petroleum soldiers" to the Persian Gulf in the vain hope that given enough M-16s and F-16s, those soldiers will be able to secure America's energy future. Instead of playing the role of the world's cop, the U.S. must work with China, Japan, and the E.U. to see that all of them play a role in ensuring the stability and security of the Persian Gulf. That requires politics and diplomacy. It also requires talking to potential foes like Iran, Syria, Cuba, and North Korea.

And yet even as America faces these myriad challenges, I remain stubbornly optimistic.

I'm optimistic because America remains a model for the world: a place where liberty, freedom, individual property, human rights, mineral rights, and the rule of law are respected. Those rights have created tremendous prosperity and opportunity. Those rights and the wealth that has been built by exercising them have created one of the longest-lived democracies in human history. America must nurture its democratic values, return to its leadership role, and engage the world by promoting human rights, liberty, knowledge, open markets, free speech, freedom *of* religion, freedom *from* religion, and the free exchange of ideas. And a critical part of that leadership role must be embracing and promoting free and open markets in energy and energy technologies. It makes absolutely no sense to politicize energy or to think that America will be more prosperous if only it builds a "wall of energy independence" around itself.

The world is growing smaller every day. There is no room on this shrinking planet for walls of any kind. Whether the issue is energy, global carbon dioxide levels, banking, communications, or the free flow of goods, people, and ideas, the world has become an interdependent organism, one that, to prosper, must accept interdependence as a fact of life.

23

EPILOGUE TO THE PAPERBACK EDITION

PARIS HILTON DECLARES "ENERGY CRISIS SOLVED!";
ETHANOL SCAM CONTRIBUTES
TO HIGHER FOOD PRICES

I finished the final edits to the manuscript of *Gusher of Lies* in December 2007. Today, about ten months later, I can discern a few trends. The first of those trends is not necessarily unexpected, but nevertheless is discouraging: The volume and frequency of the political rhetoric advocating "energy independence" has increased. And that rhetoric is coming from Republicans, Democrats, and even Paris Hilton.

The second trend is both encouraging and discouraging: It's the realization by scientists and food industry analysts that the ethanol scam is a key factor in increasing food prices. Since the publication of the hardcover edition, I've found a dozen studies that have determined that U.S. corn ethanol production is driving up food prices. The discouraging aspect of this realization is that the U.S. Congress, dominated by the Democratic Party, seems determined to ignore it and shows little willingness to roll back the ethanol juggernaut.

The other trend is the increasing disparity in gasoline and diesel prices both here in the U.S. and overseas, a disparity that is being augmented by the increasing amounts of ethanol that are flowing into the gasoline market.

Before looking at those issues, let us pause to remember that Paris Hilton is hot . . . or, at least, that's what she keeps telling us. In August 2008, Her Hotness let the world know that she, too, favors energy independence. That month, the hotel heiress/socialite/one-time porn star laid out her energy platform in a video that was released just a few days after John McCain's campaign began running a TV ad that compared Barack Obama's celebrity to that of Hilton.

Lying on a white recliner positioned on a perfectly manicured green lawn, Hilton, wearing only a bathing suit and high heels, spoke directly to the camera, saying that she, too, was planning to run for the White House. "I'm not promising change like that other guy. I'm just hot." With a version of the song "America the Beautiful" providing the musical backdrop, Hilton continued her pitch, "I want America to know that I'm, like, totally ready to lead. And now I want to present my energy policy to America." The twenty-something diva then delivered the headline: She wants to see limited offshore drilling tied to strict environmental enforcement combined with tax incentives for automakers to help them produce hybrid and electric cars. "That way, the offshore drilling carries us until the new technologies kick in, which will then create new jobs and energy independence. Energy crisis solved! I'll see you at the debates, bitches."[1]

Hilton's video provides the U.S. with yet another recipe for energy independence. Add it to a long list. The mantra of energy independence was one of the hallmarks of the 2008 presidential campaign, with both McCain and Obama promising to rid the U.S. of the scourge of foreign oil.

While the rhetoric from Hilton, McCain, and Obama bordered on the comic, the impact of the ethanol scam on food prices provides no humor at all. In the hardcover edition of Gusher of Lies, I touched briefly on the corn-ethanol–food-price question. But by the summer of 2008, it was becoming abundantly obvious that of all of the issues raised by the ethanol scam, the food price connection was likely the most pressing and the most troubling. It is true that a number of factors have had an effect on rising grain prices, including the increasing price of energy, poor harvests in other countries, the falling value of the dollar, and rising grain demand from developing countries like India and China. That said, there is simply no denying the impact that the growth of the corn ethanol business has had on food prices. And as food prices rise, so does hunger among the world's poor, as well as global unrest.

In September 2007, Corinne Alexander and Chris Hurt, agricultural economists at Purdue University, found that "about two-thirds of the increase" in food prices from 2005 to 2007 was "related to biofuels." The report also says, "Based on expected 2007 farm level crop prices, that additional food cost is estimated to be $22 billion for U.S. consumers compared to farm prices for the crops produced in 2005. A rough estimate is that about $15 billion of this increase is related to the recent surge in demand to use crops for fuel."[2]

In March 2008, a report commissioned by the Coalition for Balanced Food and Fuel Policy (a coalition based in Washington, D.C., of eight meat, dairy, and egg producers' associations) estimated that the biofuels mandates passed by Congress will cost the U.S. economy more than $100 billion from 2006 to 2009. The report declared, "The policy favoring ethanol and other biofuels over food uses of grains and other crops acts as a regressive tax on the poor." It went on to estimate that the total cost of the U.S. biofuels mandates will total some $32.8 billion in 2009, or about $108 for every American citizen.[3]

An April 8, 2008, internal report by the World Bank found that grain prices had increased by 140 percent between January 2002 and February 2008. The report declared, "This increase was caused by a confluence of factors but the most important was the large increase in biofuels production in the U.S. and E.U. Without the increase in biofuels, global wheat and maize [corn] stocks would not have declined appreciably and price increases due to other factors would have been moderate."[4] Robert Zoellick, president of the Bank, acknowledged those facts, saying that biofuels are "no doubt a significant contributor" to high food costs. And he said that "it is clearly the case that programs in Europe and the United States that have increased biofuel production have contributed to the added demand for food."[5]

In May 2008, Mark W. Rosegrant of the International Food Policy Research Institute, a Washington, D.C.–based think tank whose vision is "a world free of hunger and malnutrition," testified before the U.S. Senate on biofuels and grain prices.[6] Rosegrant said that the ethanol mandates caused the price of corn to increase by 29 percent, rice to increase by 21 percent and wheat by 22 percent.[7] Rosegrant estimated that if the global biofuels mandates were eliminated altogether, corn prices would drop by 20 percent, while sugar and wheat prices would drop by 11 percent and 8 percent, respectively, by 2010.[8] Rosegrant added, "If the current biofuel expansion continues, calorie availability in developing countries is expected to grow more slowly; and the number of malnourished children is projected to increase." He continued, "It is therefore important to find ways to keep biofuels from worsening the food-price crisis. In the short run, removal of ethanol blending mandates and subsidies and ethanol import tariffs, in the United States—together with removal of policies in Europe promoting biofuels—would contribute to lower food prices."[9]

Yet another telling report comes from the U.S. Department of Agriculture (USDA), the federal agency that has long been one of the corn ethanol sector's biggest boosters. Its leaders have repeatedly downplayed the role of ethanol in driving up food prices.[10] And yet, in July 2008, the department released a report called "Food Security Assessment, 2007," which states very clearly that the biofuels mandates are pushing up food prices. The first page of the report says:

> The persistence of higher oil prices deepens global energy security concerns and heightens the incentives to expand production of other sources of energy including biofuels. The use of food crops for producing biofuels, grow-

ing demand for food in emerging Asian and Latin American countries, and unfavorable weather in some of the largest food-exporting countries in 2006–2007 all contributed to growth in food prices in recent years.[11]

While that admission is important, the July 2008 report also reveals the growing number of people around the world who are facing food insecurity. It says that this number jumped from 849 million in 2006 to 982 million in 2007, and that it is expected to continue rising. By 2017, the number of food-insecure people is expected to hit 1.2 billion. And, says the USDA, "short-term shocks, natural as well as economic," could make the problem even worse.

Those short-term shocks are more likely because world grain stocks are at or near historic lows. In July 2008, the USDA released another report, this one called "Global Agricultural Supply and Demand: Factors Contributing to the Recent Increase in Food Commodity Prices," which showed that global grain reserves were at their lowest levels since 1970. Furthermore, those grain reserves—about 300 million tons—were less than half the reserve volumes on hand as recently as 1997.[12] And the agency expects the trend to continue. In fact, it expects that U.S. stocks of corn, wheat, and soybeans will be at near historic lows through 2016.[13]

Why are grain stocks so low? Again, let's look to the USDA for guidance. In a 2007 report, the agency plainly states that ethanol production is to blame for the lower grain reserves, saying, "Strong ethanol demand sharply lowers U.S. corn stocks in the projections."[14]

On October 7, 2008, the United Nations Food and Agriculture Organization weighed into the debate with a 138-page report called "Biofuels: Prospects, Risks and Opportunities." In the section on food, the report concludes that "rapidly growing demand for biofuel feedstocks has contributed to higher food prices, which pose an immediate threat to the food security of poor net food buyers (in value terms) in both urban and rural areas."[15]

All of these reports are essentially making the same point: The U.S. is helping to set the table for what could be a global food disaster. In

FIGURE 24 U.S. Stocks-to-use Ratios, 1980 to 2016: Corn, Wheat, and Soybeans

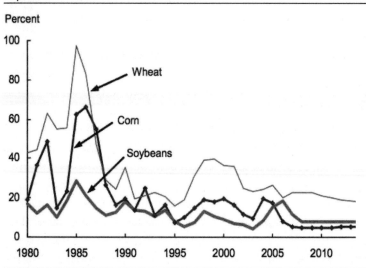

SOURCE: U.S. Department of Agriculture, "USDA Long-term Projections, February 2007," 31. Available: http://www.ers.usda.gov/publications/oce071/oce20071c.pdf.

February 2008, William Doyle, the CEO of Potash, the world's largest maker of crop fertilizers, pointed out that the paltry amount of food in storage imperils many of the world's poor. Doyle warned that if there is a major drought or other natural disaster that affects global grain supplies, "I believe you'd see famine."[16]

By using vast amounts of corn to make motor fuel, the U.S. has, in effect, militarized its food supply. The result is smaller stocks of grain and higher prices that could result in famine in countries around the world. In late 2007 and 2008, there were violent protests over food prices in Egypt, Cameroon, Ivory Coast, Mauritania, Ethiopia, Madagascar, the Philippines, and Indonesia.[17] If grain prices continue upward, these types of protests—and increasing unrest due to hunger—are likely to become more common.

Despite these many reports and the potential for devastating grain price increases, ethanol boosters continue to claim that they are not to blame. For instance, on its Web site, the Renewable Fuels Association, a trade group funded by the ethanol producers, has claimed that "corn demand for ethanol has no noticeable impact on retail food prices."[18]

In April 2008, Sean O'Hanlon, the executive director of the American Biofuels Council, when asked how biofuels are affecting food prices, replied, "They really don't." He went on to assert that "ultimately, there is more food available because of biofuels rather than less."[19]

Set America Free, the group founded in 2004 by a group of pro–Iraq War neoconservatives that includes former CIA director James Woolsey, also claims that the corn ethanol scam is not having a significant effect on food prices. Set America Free has been among the most ardent boosters of ethanol as a way to reduce America's oil import bill. And as part of its continuing attacks on foreign oil in general, and the Saudis in particular, the group has gone out of its way to try to convince citizens that corn ethanol is a force for good. For instance, in a May 6, 2008, editorial in the *Chicago Tribune,* titled "Food vs. Fuel, a Global Myth," Set America Free founder, Gal Luft, and his fellow traveler, Robert Zubrin, a vituperative ideologue who advocates colonizing Mars, declared that "farm commodity prices have almost no effect on retail prices." The two went on to claim that if only more automobiles were manufactured as "flex fuel"—that is, able to burn fuel mixtures containing 85 percent ethanol—then oil would have to compete for its share of the motor fuel market against alcohol fuels made from food crops, weeds, crop residue, and other materials. That competition, they insist, would help reduce terrorism by bankrupting the petrostates. The pair concluded their May 6 screed by saying that "rather than shut down biofuel programs, we need to radically augment them, to the point where we can take down" the Organization of the Petroleum Exporting Countries (OPEC).[20]

Zubrin, Luft, and the farm lobby continue to use claims about the need for "energy independence" and the dangers of foreign oil as justifications for ever more consumption of corn ethanol. And they are

doing so even though the U.S. now has some 200 corn ethanol distilleries that are gobbling up about one-third of all the corn grown in the
U.S., about 4.1 billion bushels per year. That's 2 times the volume of
corn consumed by the American ethanol industry in 2006, 7 times the
quantity used in 2000, and nearly 13 times the amount used in 1990.[21]

Need another comparison? That 4.1 billion bushels of corn being
used to make ethanol is more than 2 times as much corn as was produced by the entire European Union in 2006 and more than 5 times as
much as was raised in Mexico.[22]

IT'S THE DIESEL, STUPID! WHY CORN ETHANOL HASN'T (AND WON'T) CUT U.S. OIL IMPORTS

H. L. Mencken once remarked that there is always a "well-known solution to every human problem—neat, plausible, and wrong."[23]

That quote comes to mind when one considers the claims put forward by Zubrin, Luft, and their ilk who claim that the best way to cut
American oil imports, and thereby impoverish the petrostates (and in
theory, reduce terrorism), is to require automakers to manufacture
"flex-fuel" cars.

Their rationale is simple: Using more ethanol from corn or other
biomass, as well as methanol from coal or other sources, will create
competition in the motor fuel market and thereby depose oil from its
primacy as the main transportation fuel. Once that is done, oil is no
longer a strategic commodity, the price falls, the petrostates are bankrupted, and a newly energy-independent U.S. zooms back to its position as the world's undisputed sole superpower. Their rhetoric is so
attractive that several members of Congress have introduced legislation that would require the production of flex-fuel cars.

It's a simple idea that betrays a near-complete ignorance of the
world petroleum business. The ethanol producers and the flex-fuel-
car advocates are wrong because their fuel displaces only one of the
myriad products that are derived from a barrel of crude oil. That
displacement achieves next to nothing because when it is refined, a
barrel of crude yields several different "cuts" that range from light

products like butane to heavy products like asphalt.[24] Even the best-quality barrel of crude (42 gallons) yields only about 20 gallons of gasoline.[25] Furthermore, certain types of crude oil—like light sweet, which has low viscosity and contains low amounts of sulfur—are better suited to gasoline or diesel fuel production than heavy sour crude, which is thick and sulfurous. The overall point is that even the most advanced refineries cannot produce just one product from a barrel of crude; they must produce several, and the market value of those various cuts is constantly changing.

And therein lies the problem: There's very little growth in gasoline demand. Meanwhile, demand for other cuts of the crude barrel is booming. In short, the corn ethanol producers are making the wrong type of fuel at the wrong time. They are producing fuel that displaces gasoline at a time when gasoline demand both in the U.S. and globally is essentially flat. Meanwhile, demand for the segment of the crude barrel known as middle distillates—primarily diesel fuel and jet fuel—is growing rapidly. And ethanol cannot replace diesel or jet fuel, the liquids that propel the vast majority of our commercial transportation machinery.

Between 1997 and 2007, diesel consumption in the U.S. jumped by 22 percent, while gasoline demand increased by just 15 percent.[26] In June 2008, the Energy Information Administration (EIA) released its Annual Energy Outlook, which expects domestic demand for diesel fuel to grow about four times faster than that for gasoline through 2015. Further out, toward 2030, diesel demand is expected to increase more than six times faster than that for gasoline. By 2030, the EIA expects diesel consumption to rise by nearly 52 percent over 2006 consumption levels, while gasoline use will increase by just 8 percent.[27]

In July 2008, the Paris-based International Energy Agency (IEA) released its medium-term oil market report, which said that the "bulk of oil demand growth" worldwide "will be concentrated in transportation fuels." It went on to state that while global gasoline demand is growing, "distillates (jet fuel, kerosene, diesel, and other gasoil) have become—and will remain—the main growth drivers of world oil demand."[28] Between 2007 and 2013, the IEA expects distillate demand to nearly double, while global gasoline demand will grow only slightly.[29] The slower growth in the gasoline market is due to several factors, among them the maturation of the U.S. automobile market, which is not growing as fast as markets in Asia, and the increasing efficiency of the U.S. auto fleet. At the same time, the "dieselization" of the European automobile market continues. In some western European countries, more than half the new cars sold are powered by diesel engines, and that trend is expected to continue. This increasing demand for diesel, combined with a global lack of refineries that can produce the type of low-sulfur diesel that is now mandated in the U.S. and Europe, means that diesel will continue selling for a premium relative to gasoline, and that price differential will likely persist for a decade or more to come.

That increasing diesel demand (and the increasing value of diesel fuel) means that U.S. refineries are buying more foreign crude, not less. It's a bitter fact, given that cutting dependence on foreign oil has been cited over and over as the justification for the corn ethanol mandates as well as continued federal research funding for the mirage of cellulosic ethanol.

All of this complicates matters for refiners, who are scrambling to produce as much diesel as they can to meet the increasing global demand. But because their refineries cannot turn all of the crude they run in their refineries into diesel, they are also producing more gasoline, even though profits on gasoline may be small or nonexistent. As an executive at a large domestic oil refiner (who asked that his name and company not be named) explained to me, "Gasoline is being thrown into the world market as a diesel by-product."

In other words, ethanol is doing absolutely nothing to reduce overall U.S. oil consumption or imports because refiners are having to buy the same amount of crude (or more) in order to meet the demand for products other than gasoline, that is, jet fuel, diesel fuel, fuel oil, asphalt, and so on. Proof can be found in the data between 2000 and 2008, the period during which U.S. ethanol production capacity more than quadrupled. By July 2008—according to the Renewable Fuels

Association—domestic ethanol companies were producing more than 600,000 barrels of ethanol per day.[30]

Given that production capacity, the ethanol boosters should be able to point to data that show a decrease—or at a minimum, a reduction—of U.S. oil imports. They cannot. Between mid-2000 and mid-2008, domestic oil consumption was essentially flat, at about 19.5 million barrels per day.[31] Over that same time period, domestic crude oil production fell by about 600,000 barrels per day—by coincidence about the same volume as current domestic ethanol production.[32] And yet, between mid-2000 and mid-2008, oil imports jumped by about 1.4 million barrels per day. In July 2000, the U.S. was importing about 11.6 million barrels of crude oil and petroleum products per day. By July 2008, imports had increased to about 13 million barrels per day.[33]

The punch line here is obvious: The corn ethanol scam has done little or nothing to reduce U.S. oil use or oil imports. Unless or until inventors can come up with a substance that can replace essentially all of the types of fuel that are refined from a barrel of crude oil—from gasoline to naphtha and diesel to jet fuel—the U.S. is likely to continue using oil as its primary transportation fuel for many years to come. And that will be true no matter how much corn gets burned up in America's delusional quest for "energy independence."

While the continued calls for energy independence and the increased use of corn ethanol are distressing, 2008 also brought some positive news on the energy front. Several automakers announced plans to develop and sell electric cars in the U.S. market by 2010. Perhaps even more important was the success of domestic natural gas producers. In mid-2008, the EIA reported that between the first quarter of 2007 and the first quarter of 2008, U.S. gas production increased by a whopping 9 percent.[34] And in October 2008, the EIA announced that in 2007 U.S. gas reserves increased by 13 percent. (Oil reserves increased by just 2 percent.)[35] The huge increases in both production and reserves is largely due to breakthroughs in tapping natural gas deposits in shale deposits like the Barnett Shale in Texas, the Woodford Shale in Oklahoma, the Fayetteville Shale in Arkansas, and other shale deposits around the country. The surge in gas production and reserves helped drive U.S. natural gas prices downward in the second half of 2008. But over the coming years, the U.S. will enjoy natural gas prices that are far below those of other countries—particularly Asian countries—and that should provide a strategic economic advantage to America. Further, America's cheap natural gas is convincing more fleet owners to convert their trucks to using natural gas as motor fuel. While the idea of natural gas vehicles is not new, the push to increase the use of natural gas in the terrestrial transportation sector gained traction in 2008, thanks to surging diesel fuel prices and a high-profile campaign paid for by Dallas billionaire T. Boone Pickens. In addition, nuclear power appears to be on the rebound in the U.S. And while most of the focus is on the big nuclear reactors with an output of 1,000 megawatts or more, perhaps the most promising work is happening in the field of micro reactors. Since finishing the hardback edition of *Gusher of Lies*, I've learned of two U.S. companies, Hyperion Power Generation and NuScale Power, that are developing reactors that will produce about 30 megawatts of electric power. That's a fraction of the size of the reactors now being used by nuclear utilities, but they would be perfect for use in remote areas, like rural Canada and rural Alaska, or in association with large-scale mining operations like the ones under way in Canada that are extracting petroleum from tar sands. Hyperion says its unit is about the size of a hot tub and could be fit onto the back of a semitrailer truck. The company also claims that the fuel used by its reactor does not pose a proliferation risk because it uses uranium hydride, a substance that requires intense refining before it can be turned into a useful nuclear weapon. And the company foresees a huge potential market. Hyperion says that if it gets licensed, it could build 4,000 of these small reactors in its first decade of operation. Both NuScale and Hyperion are planning to submit license applications to the Nuclear Regulatory Commission (NRC) in 2009. If the applications are approved in a timely manner (the NRC is underfunded and undermanned), these new devices could provide a real boost to the supply and stability of the U.S. power grid. They could also be a boon to the electrification of the automotive sector as they would be able to provide relatively inexpensive base-load power.

There are other positive developments, including growing interest in producing biofuel from algae. In September 2008, Sapphire Energy, a U.S.-based startup, announced it had raised a total of more than $100 million from various investors (that group includes Cascade Investment, an investment holding company owned by Microsoft founder Bill Gates) to further the development of what it calls "green crude."[36] And in October 2008, the Carbon Trust, a publicly funded company in the United Kingdom, announced a major effort to commercialize the use of algae biofuel by 2020.[37]

Meanwhile, numerous battery companies, both in the U.S. and abroad, are continuing to develop better batteries. All of those developments leave me optimistic about our potential to meet the energy challenges that face the U.S. and the rest of the world.

That said, the U.S. is facing numerous hard choices in the years ahead. The costly financial bailout of 2008, the crippling federal debt, an ailing economy, a decaying infrastructure, and the huge costs of the Second Iraq War have left America with very little latitude with regard to its energy options. Furthermore, the collapse in oil prices—which hit $145 in July 2008 and then fell to about $60 by late October 2008—is likely to slow, or even stop, investment in renewable and alternative energy sources.

The wise course for the U.S. lies in accepting the reality of energy interdependence. It also should include increased domestic oil and gas production through expanded drilling, both onshore and offshore; and an embrace of nuclear power, which can help provide the electricity that will be needed to charge the batteries that will propel the growing numbers of electric cars likely to be plying the roads in years to come. It should also encourage the use of natural gas as a motor fuel. If the U.S. can move rapidly on all of those fronts, then there will be even more reasons to be optimistic.

APPENDIX A

U.S. Imports of Strategic Mineral Commodities, Their Uses, and Key Suppliers

Commodity	Percent Imported	Key Uses	Primary Suppliers
Arsenic	100	pesticides, wood preservatives	China, Chile, Morocco, Mexico
Asbestos	100	heat and acoustic insulation, fire proofing, roofing and flooring	Canada
Indium	100	electronic displays and devices, glass coating	China, Canada, Japan, France
Manganese	100	component of steel and iron, fertilizer, colorant	South Africa, Gabon, Australia, France
Strontium	100	nuclear fuel, ceramics and glass (TV faceplates)	Mexico, Germany
Yttrium	100	ceramics and glass (TVs), microwaves, jewelry	China, Japan, Austria, Netherlands
Vanadium	100	steel additive, auto parts, ceramics, dyes	Czech Republic, South Africa, Canada, China
Bauxite and Alumina	100	aluminum metal, chemical products, abrasives, transportation, building construction, packaging, electrical	Australia, Jamaica, Guinea, Suriname

continues

continued

Commodity	Percent Imported	Key Uses	Primary Suppliers
Columbium	100	additive in steelmaking, superalloys, combustion equipment, jet engines, superconductors	Brazil, Canada, Estonia, Germany
Fluorspar	100	hydrofluoric acid (electroplating, stainless steel, refrigerant, plastics), aluminum fluoride (aluminum smelting, ceramics and glass, steel)	China, South Africa, Mexico
Graphite	100	lubricant, refractory applications, linings, molds	China, Mexico, Canada, Brazil
Mica	100	electronic insulators (vacuum tubes), paint, cement, plastics, roofing, rubber, welding rods	India, Belgium, China, Germany
Quartz Crystal	100	semiprecious gemstones, pressure gauges, prisms, spectrographic lenses, glass, paints, abrasives, refractory, precision instruments	Brazil, Germany, Madagascar
Rare Earths	100	catalysts, metallurgical additives, ceramics and polishing compounds, permanent magnets, phosphors	China, France, Japan, Estonia
Rubidium	100	medical and electronic applications, laboratory studies	Canada
Thallium	100	electronics (gamma radiation detection equipment, infrared radiation detection and transmission equipment), glass, catalysts, medical equipment	Belgium, France, Russia, United Kingdom
Thorium	100	coatings (light bulbs, lab equipment), optical lenses, alloys (aerospace industry)	France

Commodity	Percent Imported	Key Uses	Primary Suppliers
Gallium	99	specialized electrical applications, lasers, photo detectors, LEDs, solar cells, integrated circuits, semiconductors, thermometers, mirrors	France, China, Russia, Kazakhstan
Platinum	91	automobile catalytic converters, jewelry, electronics	South Africa, United Kingdom, Germany, Canada, Russia
Bismuth	90	alloys, pharmaceuticals, chemicals, ceramics, paints, catalysts, medicine	Belgium, Mexico, China, United Kingdom
Tin	88	electroplating, utensils, electronics, pharmaceuticals, ammunition	Peru, China, Bolivia, Brazil
Antimony	85	alloys (storage batteries and cable sheaths), bearing metal, type metal, collapsible equipment, semiconductors	China, Mexico, South Africa, Belgium
Diamond	85	jewelry, abrasives, steel drill bits, drilling machinery	Ireland, Switzerland, United Kingdom, Russia
Titanium	85	airplanes, pigment (paint, paper, plastics)	Kazakhstan, Japan, Russia
Palladium	81	automobile catalytic converters, electronics, jewelry, dentistry	Russia, South Africa, United Kingdom, Belgium, Germany
Tantalum	80	metal powder for electronic components, alloys (cemented carbide tools for metal working equipment, jet engines)	Australia, Kazakhstan, Canada, China

continues

continued

Commodity	Percent Imported	Key Uses	Primary Suppliers
Barite	79	oil drilling, cement, filler, medical uses, industrial products	China, India
Rhenium	79	alloys (jet engines, heating elements, mass spectrographs, electrical contacts, electromagnets, semiconductors), superconductors	Chile, Kazakhstan, Mexico
Cobalt	76	aircraft engines, magnets, alloys, steel, catalysts	Finland, Norway, Russia, Canada
Iodine	74	medicine, photography, nutrition	Chile, Japan, Russia
Tungsten	73	steel, metal alloys, electronics	China, Canada
Chromium	72	steel, catalysts	South Africa, Kazakhstan, Zimbabwe, Russia
Potash	70	glass, potassium silicate, printing inks, soaps, food additive, fertilizer	Canada, Belarus, Russia, Germany
Magnesium Metal	68	photography, airplane and missile construction, pyrotechnics	Canada, Russia, China, Israel
Titanium Mineral Concentrates	65	alloys (jet engines, missiles, spacecraft), industrial uses (chemicals, petrochemicals, desalination plants, paper), automotive, medical equipment	South Africa, Australia, Canada, Ukraine
Oil/Oil Products	60	transportation, lubricants, sealants, drugs, plastics, solvents, adhesives, fertilizer, packaging	Canada, Mexico, Venezuela, Saudi Arabia, Nigeria
Silicon	56	concrete, brick, glass, electronics	South Africa, Norway, Brazil, Russia

Commodity	Percent Imported	Key Uses	Primary Suppliers
Peat	56	agriculture, horticulture, potting soil, absorbent	Canada
Zinc	56	alloys, metals, anticorrosion coatings, construction, brass	Canada, Mexico, Peru
Beryllium	55	alloys, structural reinforcement, X-rays, reflector in nuclear reactors	Kazakhstan, Japan, Brazil, Spain
Silver	54	appliances, electrical conductors, jewelry, metals, photography	Mexico, Canada, United Kingdom, Peru
Nickel	49	engineering, metals, transportation, electronics	Canada, Russia, Norway, Australia
Magnesium Compounds	48	refractory material for producing iron, steel, nonferrous metals, glass, cement	China, Australia, Canada, Austria
Copper	43	circuits, electrical conductors, electronics, heating, automotive	Canada, Chile, Peru, Mexico

SOURCES: U.S..Geological Survey, "The U.S. Geological Survey Mineral Resources Program Five-Year Plan, 2006–2010," updated November 10, 2005, 3. Available: http://minerals.usgs.gov/plan/2006-2010/mrp-plan-2006-2010.pdf. Usage data were obtained primarily from the Mineral Information Institute.

APPENDIX B

The Set America Free Manifesto, Released on September 27, 2004, at the National Press Club, Washington, D.C.*

AN OPEN LETTER TO THE AMERICAN PEOPLE

For decades, the goal of reducing the Nation's dependence upon foreign energy sources has been a matter on which virtually all Americans could agree. Unfortunately, differences about how best to accomplish that goal, with what means, how rapidly and at what cost to taxpayers and consumers have, to date, precluded the sort of progress that might have been expected before now.

Today, we can no longer afford to allow such differences to postpone urgent action on national energy independence. After all, we now confront what might be called a "perfect storm" of strategic, economic and environmental conditions that, properly understood, demand that we affect [sic] over the next four years a dramatic reduction in the quantities of oil imported from unstable and hostile regions of the world.

America consumes a quarter of the world's oil supply while holding a mere 3% of global oil reserves. It is therefore forced to import over 60% of its oil, and this dependency is growing. Since most of the world's oil is controlled by countries that are unstable or at odds with the United States this dependency is a matter of national security.

At the **strategic level**, it is dangerous to be buying billions of dollars worth of oil from nations that are sponsors of or allied with radical Islamists who

---

*Note that all boldfaced text is the same as in the original document.

313

foment hatred against the United States. The petrodollars we provide such nations contribute materially to the terrorist threats we face. In time of war, it is imperative that our national expenditures on energy be redirected away from those who use them against us.

Even if the underwriting of terror were not such a concern, our present dependency creates unacceptable vulnerabilities. In Iraq and Saudi Arabia, America's enemies have demonstrated that they can advance their strategic objective of inflicting damage on the United States, its interests and economy simply by attacking critical overseas oil infrastructures and personnel. These targets are readily found not only in the Mideast but in other regions to which Islamists have ready access (e.g., the Caspian Basin and Africa). To date, such attacks have been relatively minor and their damage easily repaired. Over time, they are sure to become more sophisticated and their destructive effects will be far more difficult, costly and time-consuming to undo.

Another strategic factor is China's burgeoning demand for oil. Last year, China's oil imports were up 30% from the previous year, making it the world's No. 2 petroleum user after the United States. The bipartisan, congressionally mandated U.S.-China Economic and Security Review Commission reported that: "China's large and rapidly growing demand for oil is putting pressure on global oil supplies. This pressure is likely to increase in the future, with serious implications for U.S. oil prices and supplies."

Oil dependence has considerable **economic implications**. Shrinking supply and rising demand translate into higher costs. Both American consumers and the U.S. economy are already suffering from the cumulative effect of recent increases in gas prices. Even now, fully one-quarter of the U.S. trade deficit is associated with oil imports. By some estimates, we lose 27,000 jobs for every billion dollars of additional oil imports. Serious domestic and global economic dislocation would almost certainly attend still-higher costs for imported petroleum and/or disruption of supply.

Finally, **environmental considerations** argue for action to reduce imports of foreign oil. While experts and policy-makers disagree about the contribution the burning of fossil fuels is making to the planet's temperatures, it is certainly desirable to find ways to obtain energy while minimizing the production of greenhouse gases and other pollutants.

The combined effects of this "perfect storm" require concerted action, at last, aimed at reducing the Nation's reliance on imported oil from hostile or unstable sources and the world's dependence on oil at large. Fortunately, with appropriate vision and leadership, we can make major strides in this direction by exploiting currently available technologies and infrastructures to greatly diminish oil consumption in the transportation sector, which accounts for two thirds of our oil consumption.

The . . . **Blueprint for Energy Security: "Set America Free"** spells out practical ways in which real progress on "fuel choice" can be made over the next four years and beyond. To be sure, full market transformation will take a longer time. In the case of the transportation sector, it may require 15–20 years. That is why it is imperative to begin the process without delay.

We call upon America's leaders to pledge to adopt this Blueprint, and embark, along with our democratic allies, on a multilateral initiative to encourage reduced dependence on petroleum. In so doing, they can reasonably promise to: deny adversaries the wherewithal they use to harm us; protect our quality of life and economy against the effects of cuts in foreign energy supplies and rising costs; and reduce by as much as 50% emissions of undesirable pollutants. In light of the "perfect storm" now at hand, we simply can afford to do no less.

Milton Copulos
National Defense Council Foundation

Frank Gaffney
Center for Security Policy

Bill Holmberg
American Council on
Renewable Energy

Anne Korin
Institute for the
Analysis of Global Security

Gal Luft
Institute for the Analysis of
Global Security

Cliff May
Foundation for the
Defense of Democracies

Robert C. McFarlane
Former National Security Advisor

R. James Woolsey
Co-Chairman, Committee on
the Present Danger

Meyrav Wurmser
Hudson Institute

APPENDIX C

A Survey of Fuel Sources:
A Comparison of Their Land Use Requirements,
Water Use, Energy Ratios, and Carbon Emissions

Notes:
[a] M = million, B = billion.
[b] This includes driving cycle. Carbon emissions are calculated from the chemical composition of the various fuels and the appropriate stoichiometric combustion equations with the consumption of CO_2 by photosynthesis not included so that all fuels are compared on an equivalent basis. To calculate CO_2 emissions from the carbon number given, multiply by 3.7.
[c] Algaculture is added to represent new results that bear on this paper but is not presented herein because of space limitations.

SOURCE: Jan F. Kreider and Peter S. Curtiss, "Comprehensive Evaluation of Impacts From Potential, Future Automotive Fuel Replacements," Proceedings of Energy Sustainability 2007, June 27–30, 2007, 11.

| Fuel Source | Transportation energy displacement | Land use | | | | Water use (gallons) | | Energy ratio | Carbon emissions[b] |
		Acres[a]	Fraction of U.S. cropland	Gallons of fuel per acre	MMBtu of fuel per acre	Per gallon of fuel	Per MMBtu of fuel	Btu input per Btu of fuel output	Lb. per MMBtu of fuel
Conventional gasoline	0–100%	tens of thousands	very low	-	-	5	45	0.05	60
Conventional diesel	0–100%	tens of thousands	very low	-	-	10	80	0.09	60
Corn-based ethanol	10%	65 M	20%	370	28	170	2,200	0.98	95
	25%	160 M	51%	370	28	180	2,300	0.98	95
	50%	337 M	103%	360	28	220	2,900	0.98	95
Cellulosic ethanol	10%	46 M	15%	515	39	146	1,900	0.92	90
	25%	112 M	35%	515	39	146	1,900	0.92	90
	50%	228 M	72%	510	39	149	1,900	0.92	90
Soybean biodiesel	10%	253 M	80%	57	7	900	6,900	0.76	50–60
	25%	380 M	120%	57	7	900	6,900	0.76	50–60
	50%	1.2 B	390%	57	7	900	6,900	0.76	50–60
Algaculture[c]	10%	2.5 M	< 1%	6,000	800	50	400	0.2	absorbs
	25%	6.5 M	2%	6,000	800	50	400	0.2	waste
	50%	13 M	4%	6,000	800	50	400	0.2	power plant CO_2

NOTES

INTRODUCTION

1. Richard Nixon, State of the Union address, January 30, 1974. Available: http://www.thisnation.com/library/sotu/1974rn.html.

2. Gerald Ford, State of the Union address, January 15, 1975. Available: http://www.ford.utexas.edu/LIBRARY/SPEECHES/750028.htm.

3. Jimmy Carter, televised speech on energy policy, April 18, 1977. Available: http://www.pbs.org/wgbh/amex/carter/filmmore/ps_energy.html.

4. Greenpeace is perhaps the most insistent of the environmental groups regarding energy independence. This 2004 statement is fairly representative: http://www.greenpeace.org/international/campaigns/no-war/war-on-iraq/it-s-about-oil. For Worldwatch, see its press release after George W. Bush's 2007 State of the Union speech, which talks about "increased energy independence." Available: http://www.worldwatch.org/node/4873.

5. See any number of presentations by Lovins on energy independence. One sample: his presentation before the U.S. Senate Committee on Energy and Natural Resources on March 7, 2006. Available: http://energy.senate.gov/public/index.cfm?FuseAction=Hearings.Testimony&Hearing_ID=1534&Witness_ID=4345. Or see *Winning the Energy Endgame,* by Lovins et al., 228, discussing the final push toward "total energy independence" and the move to the hydrogen economy.

6. National Apollo Alliance Steering Committee statement. Available: http://www.apolloalliance.org/about_the_alliance/who_we_are/steeringcommittee.cfm.

7. At approximately 1:32 into the movie, in a section that discusses what individuals can do to counter global warming, a text message comes onto the screen: "Reduce our dependence on foreign oil, help farmers grow alcohol fuels."

8. AMPAS data. Available: http://www.oscars.org/79academyawards/nomswins.html.

9. Barack Obama, "Energy Security Is National Security," Remarks of Senator Barack Obama to the Governor's Ethanol Coalition, February 28, 2006. Available: http://obama.senate.gov/speech/060228-energy_security_is_national_security/index.html.

10. Original video at www.votehillary.org. See also, http://www.washington post.com/wp-dyn/content/article/2007/01/20/AR2007012000426.html.

11. JohnEdwards.com, "Achieving Energy Independence and Stopping Global Warming through a New Energy Economy," undated. Available: http://johnedwards .com/about/issues/energy/new-energy-economy.

12. *New York Times,* "Energy Time: It's Not about Something for Everyone," January 16, 2007.

13. Shailagh Murray, "Ethanol Undergoes Evolution as Political Issue," *Washington Post,* March 13, 2007, A06. Available: http://www.washingtonpost.com/wp-dyn/content/article/2007/03/12/AR2007031201722_pf.html.

14. Richard Perez-Pena, "Giuliani Focuses on Energy," *The Caucus: Political Blogging from the New York Times,* March 14, 2007. Available: http://thecaucus .blogs.nytimes.com/2007/03/14/giuliani-focuses-on-energy.

15. Mitt Romney, e-mail message, "Join Mitt," sent through jpostoffers@ jpostmail.com, April 26, 2007.

16. In 2005, Syria exported about 155,000 barrels of oil per day. Energy Information Administration (EIA) data. Available: http://www.eia.doe.gov/emeu/cabs/East_Med/Full.html.

17. States News Service, "Pelosi: Democrats' New Direction Will Restore Economic Security for All Americans," October 5, 2006.

18. Available: http://www.kicktheoilhabit.org/phase1/index.php.

19. Robert Redford, "Redford: Kicking the Oil Habit," CNN.com, May 30, 2006. Available: http://www.cnn.com/2006/US/05/30/redford.oil/index.html.

20. CNN.com, "A Lively Discussion on Rising Oil Prices," May 17, 2006. Available: http://transcripts.cnn.com/TRANSCRIPTS/0605/17/lkl.01.html.

21. Andy Grove, "Thinking Strategically," *Wall Street Journal,* January 22, 2007, A15.

22. PR Web, "The 'Green Cowboy' S. David Freeman on Winning Our Energy Independence by Using Renewable Resources," August 15, 2007. Available: http://fe24.news.sp1.yahoo.com/s/prweb/20070815/bs_prweb/prweb546992.

23. Yale Center for Environmental Law and Policy, 2007 Environment survey. Available: http://www.yale.edu/envirocenter/YaleEnvironmentalPoll2007Key findings.pdf.

24. UPI, "Americans Want Energy Action, Poll Says," April 17, 2007. Available: http://www.upi.com/Energy/Briefing/2007/04/17/americans_want_energy_action _poll_says.

25. Jane Harman, "A Bright Idea for America's Energy Future," *Huffington Post*, March 15, 2007. Available: http://www.huffingtonpost.com/rep-jane-harman/a-bright-idea-for-america_b_43519.html.

26. http://www.infoplease.com/ipa/A0922041.html.

27. Organization of Arab Petroleum Exporting Countries (OPEC), *Annual Statistical Report 2006*, 75. Available: http://www.oapecorg.org/images/A%20S%20R%202006.pdf.

28. Nazila Fathi and Jad Mouawad, "Unrest Grows amid Gas Rationing in Iran," *New York Times*, June 29, 2007. According to this story, Iran imports gasoline from 16 countries. Iran has been importing natural gas from Turkmenistan since the late 1990s. In 2008, those imports will likely be about 1.3 billion cubic feet of natural gas per day. The fuel will be used to meet demand in northern Iran. For more, see, David Wood, Saeid Mokhatab, and Michael J. Economides, "Iran Stuck in Neutral," *Energy Tribune*, December 2006, 19.

29. EIA oil reserve data for Saudi Arabia available: http://www.eia.doe.gov/emeu/cabs/saudi.html. EIA oil reserve data for Iran available: http://www.eia.doe.gov/emeu/cabs/Iran/Oil.html. EIA natural gas data for Iran available: http://www.eia.doe.gov/emeu/cabs/Iran/NaturalGas.html.

30. Council on Foreign Relations, "National Security Consequences of U.S. Oil Dependency," October 2006, 4. Available: http://www.cfr.org/content/publications/attachments/EnergyTFR.pdf.

31. A June 2007 survey done by Harris Interactive for the American Petroleum Institute found that only 9 percent of the respondents named Canada as America's biggest supplier of oil for the year 2006. For more on this, see Robert Rapier, "America's Energy IQ," R-Squared Energy Blog, June 29, 2007. Available: http://i-r-squared.blogspot.com/2007/06/americas-energy-iq.html#links. For the results of the entire survey, see: http://www.energytomorrow.org/energy_issues/energy_iq/energy_iq_survey.html.

32. EIA crude import data available: http://tonto.eia.doe.gov/dnav/pet/pet_move_impcus_a2_nus_epc0_im0_mbbl_a.htm. EIA data for jet fuel available: http://tonto.eia.doe.gov/dnav/pet/pet_move_impcus_a2_nus_EPJK_im0_mbbl_a.htm. EIA data for finished motor gasoline available: http://tonto.eia.doe.gov/dnav/pet/pet_move_impcus_a2_nus_epm0f_im0_mbbl_a.htm.

33. EIA coal data available: http://www.eia.doe.gov/cneaf/coal/quarterly/html/t18p01p1.html. For gas imports, EIA data available: http://tonto.eia.doe.gov/dnav/ng/ng_move_impc_s1_a.htm.

34. EIA data available: http://www.eia.doe.gov/cneaf/electricity/epa/epat6p3.html.

35. Information from 2006, EIA data available: http://www.eia.doe.gov/cneaf/nuclear/umar/table3.html.

36. The U.S. imports about 20 percent of its steel. See Daniel Ikenson, "It's Not Just Oil," Cato Institute, September 9, 2005. Available: http://www.freetrade.org/node/311.

CHAPTER 1

1. EIA data. Available: http://tonto.eia.doe.gov/dnav/pet/pet_move_impcus _a2_nus_ep00_im0_mbblpd_a.htm.

2. Ivan Eland, "Have 1,000 U.S. Souls Died for Oil?" Independent Institute, September 13, 2004. Available: http://www.independent.org/newsroom/article .asp?id=1362.

3. Ibid.

4. U.S. Geological Survey, "The U.S. Geological Survey Mineral Resources Program Five-Year Plan, 2006–2010," updated November 10, 2005, 3. Available: http://minerals.usgs.gov/plan/2006–2010/mrp-plan-2006-2010.pdf. The list of mineral commodities for which the U.S. is 100 percent dependent on imports includes arsenic, asbestos, columbium, fluorspar, graphite, indium, mica (natural sheet), quartz crystal (industrial use), rare earths, rubidium, thallium, thorium, and vanadium.

5. EIA data. Available: http://www.eia.doe.gov/emeu/steo/pub/fsheets/real _prices.html. For an Excel spreadsheet with the gasoline data from 1919 to 2008, see: http://www.eia.doe.gov/emeu/steo/pub/fsheets/real_prices.xls.

6. U.S. Census Bureau, Historical Statistics of the United States, "Series Q 148–162, Motor-Vehicle Factory Sales and Registrations, and Motor-Fuel Usage: 1900 to 1970," 716.

7. EIA, "Basic Petroleum Statistics." Available: http://www.eia.doe.gov/neic/ quickfacts/quickoil.html.

8. EIA data. Available: http://tonto.eia.doe.gov/dnav/pet/hist/mg_rco_usw .htm. Note that the $2.75 is the nominal price, not inflation-adjusted.

9. EIA data. Available: http://www.eia.doe.gov/emeu/steo/pub/fsheets/real _prices.html.

10. EIA data. Available: http://tonto.eia.doe.gov/dnav/pet/hist/mcrntus2a.htm. Note that no data are available for 1916 through 1919.

11. It's worth noting that Ford's Model T got 25 miles per gallon. That's better than the current fleet of vehicles and higher than current Corporate Average Fuel Economy (CAFE) standards.

12. EIA data. Available: http://tonto.eia.doe.gov/dnav/pet/hist/mcrntus2a.htm.

13. For 1913 GDP data see: http://eh.net/hmit/gdp/gdp_answer.php?CHK nominalGDP=on&year1=1913&year2=.

14. U.S. Census Bureau, Historical Statistics of the United States, "Series Q 148–162, Motor-Vehicle Factory Sales and Registrations, and Motor-Fuel Usage: 1900 to 1970," 716.

15. Bureau of Transportation Statistics data. Available: http://www.bts.gov/publications/national_transportation_statistics/html/table_01_11.html.

16. According to the EIA, retail electricity in the U.S. cost $0.086 per kilowatt-hour in 1960. By 2005, it had fallen to $0.072. Available: http://www.eia.doe.gov/emeu/aer/txt/ptb0810.html.

17. http://www.eia.doe.gov/emeu/aer/txt/ptb0708.html.

CHAPTER 2

1. Philip J. Deutch, "Energy Independence," *Foreign Policy*, November/December 2005, 20.

2. CBS News, "Transcript of Barack Obama's Speech," February 10, 2007. Available: http://www.cbsnews.com/stories/2007/02/10/politics/printable2458099.shtml.

3. Actual U.S. total is 2.9 gallons/capita/day. In 2006, according to the EIA, the U.S. consumed about 21 million barrels, or 882 million gallons, of oil per day. See: http://www.eia.doe.gov/emeu/steo/pub/a1tab.html. U.S. population is 301 million; see: https://www.cia.gov/library/publications/the-world-factbook/geos/us.html#People.

4. Available: http://money.cnn.com/magazines/fortune/mostadmired/2007/index.html.

5. The highest-ranking energy company was BP, which was ranked 23rd. Exxon Mobil came in at number 25. http://money.cnn.com/magazines/fortune/globalmostadmired/top50/index.html.

6. *NASCAR Nation* is the title of a book.

7. Karlyn Bowman, "The Federal Government: Losing Public Support," *Roll Call*, October 5, 2006.

8. Stan Greenberg, Amy Gershkoff, and James Carville, "Re Meltdown II," Democracy Corps, October 19, 2006. Available: http://www.democracycorps.com/reports/analyses/Democracy_Corps_October_19_2006_Memo.pdf.

9. Thomas L. Friedman, "The Energy Mandate," *New York Times*, October 13, 2006, 27.

10. Greenberg et al., op. cit.

11. Some estimates rank Venezuela and/or Canada ahead of Iraq, but those estimates are counting unconventional oil. In the case of Venezuela, that estimate is counting that country's deposits of extra-heavy crude. In Canada, the higher estimates include tar sands.

12. John Roberts, "Oil and the Iraq War of 2003," International Research Center for Energy and Economic Development, 2003, 2.

13. Michael R. Gordon and Bernard F. Trainor, *Cobra II*, 192.

14. Robert Bryce, *Cronies*, 239.

15. Ibid., 166.

16. Peter Galbraith, testimony before the Senate Foreign Affairs Committee, January 11, 2007.

17. John F. Burns and Kirk Semple, "U.S. Finds Iraq Insurgency Has Funds to Sustain Itself," *New York Times,* November 26, 2006.

18. CNN.com, "Transcript: Bush's News Conference," October 11, 2006. Available: http://www.cnn.com/2006/POLITICS/10/11/bush.transcript.

19. Author interview, A. F. Alhajji, via telephone, September 15, 2006. For Alhajji's work on Iraq and oil, see www.aalhajji.com.

20. Carolyn Lochhead, "Iraq Refugee Crisis Exploding," *San Francisco Chronicle,* January 16, 2007. Available: http://www.sfgate.com/cgi-bin/article.cgi?file=/c/a/2007/01/16/MNG2MNJBIS1.DTL.

21. Available: http://www.iraqbodycount.org.

22. Dale Keiger, "The Number," *Johns Hopkins Magazine,* February 2007. Available: http://www.jhu.edu/~jhumag/0207web/number.html.

23. Available: http://icasualties.org/oif.

24. Jason Szep, "Cost of Iraq War Could Top $2 Trillion—Study," Reuters, January 9, 2006. Bilmes was also former Assistant Secretary of Commerce in the Clinton administration.

25. The CIA estimated Saudi Arabia's 2006 GDP was $366.2 billion. Available: https://www.cia.gov/library/publications/the-world-factbook/geos/sa.html #Econ.

26. Amy Belasco, "The Cost of Iraq, Afghanistan, and Other Global War on Terror Operations Since 9/11," Congressional Research Service, updated June 28, 2007, summary page. Available: http://www.fas.org/sgp/crs/natsec/RL33110.pdf.

27. http://costofwar.com/index.html.

28. Belasco, op. cit., CRS-24.

29. Gal Luft, "Oil Kamikaze," *IAGS Energy Security,* July 12, 2004. Available: http://www.iags.org/n0712041.htm. For the record of Luft's military service, see Gal Luft, "The Palestinian H-Bomb: Terror's Winning Strategy," *Foreign Affairs,* July/August 2002. Abstract available: http://www.foreignaffairs.org/20020701 facomment8514/gal-luft/the-palestinian-h-bomb-terror-s-winning-strategy.html.

30. Bruce Lawrence, *Messages to the World,* 46.

31. Ibid., 51.

32. Ibid., 73.

33. Ibid., 163.

34. Ibid., 272.

35. http://usinfo.state.gov/is/international_security/terrorism/uss_cole.html.

36. CNNMoney.com, "Oil Skyrockets More Than $2 After Saudi Attack," February 24, 2006. Available: http://money.cnn.com/2006/02/24/markets/oil_attack/ ?cnn=yes.

37. http://www.saudi-us-relations.org/articles/2006/ioi/060228-rodhan-abqaiq .html.

38. Fred Burton, "Attacks on Energy Infrastructure: Desire, Capability and Vulnerability," Stratfor, March 2, 2006.

39. Michael Slackman, "Saudis Round Up 172, Citing a Plot against Oil Rigs," *New York Times,* April 28, 2007.

40. http://www.iags.org/iraqpipelinewatch.htm.

41. Government Accountability Office, "Security Assistance: Efforts to Secure Colombia's Caño Limón-Coveñas Oil Pipeline Have Reduced Attacks, but Challenges Remain," GAO-05-971, September 2005, 15.

42. Reuters, "Colombia Rebel Attacks Kill 4, Shut Oil Pipeline," July 26, 2006.

43. Harry Levin, "Iraq Oil Line Is Target of Arab Terrorist Bands," *Chicago Daily Tribune,* July 24, 1938.

44. Robert Goralski and Russell W. Freeburg, *Oil and War,* 106.

45. *New York Times,* "Pipeline Politics a New Arab Game," January 7, 1958.

46. Seth S. King, "Conquests Strengthen Israeli Control of Resources," *New York Times,* August 20, 1967. After a brief halt in the flow of oil, the Tapline continued operations.

47. *Washington Post,* "Arab Guerrillas Blast Oil Line Held by Israel," June 1, 1969.

48. *New York Times,* "U.S. Says Raid on Oil Base Won't Hamper Air Strikes," August 6, 1965, 1.

49. Walter Rostow, memo to Lyndon Johnson, June 16, 1966.

50. *New York Times,* "Oil for U.S. Linked by Saudi to Peace," April 20, 1973, 7.

51. Kenneth S. Deffeyes, *Hubbert's Peak,* 10.

52. Kenneth Deffeyes and Peter Huber, "It's the End of Oil/Oil Is Here to Stay," *Time,* October 23, 2005. Available: http://www.time.com/time/magazine/article/ 0,9171,1122019-1,00.html.

53. Michael Klare, *Blood and Oil,* 21.

54. Ibid., "autonomy" at 202; "independence" at 25.

55. Paul Roberts, *The End of Oil,* 10.

56. Ibid., 331.

57. Matthew Simmons, *Twilight in the Desert,* 244.

58. James Howard Kunstler, *The Long Emergency,* jacket flap.

59. Ibid., 293.

60. Ibid., 82.

61. See Robert B. Semple, Jr., "The End of Oil," *New York Times,* February 28, 2006. Available: http://www.energybulletin.net/13368.html. See also Dan Neil, "End Times," *Los Angeles Times,* May 21, 2006. Available: http://www.latimes .com/features/printedition/magazine/la-tm-neil21may21,1,4476344.story ?coll=la-headlines-magazine.

62. H. Josef Hebert, "Report: Demand to Outpace Crude Supplies," Associated Press, July 16, 2007.

63. MarketWatch, "IEA Sees Oil Supply Crunch after 2010," July 9, 2007.

64. Richard Heinberg, *The Party's Over*, 105.

65. *Los Angeles Times*, "U.S. Warned Oil Shortage Due in 20 Years," August 18, 1946, 6.

66. Heinberg, op. cit., 106.

67. Robert L. Bradley, Jr., and Richard W. Fulmer, *Energy*, 81.

68. Ibid.

69. Eugene Gholz and Daryl G. Press, "Energy Alarmism: The Myths That Make Americans Worry about Oil," Cato Institute, April 5, 2007, 6. Available: http://www.cato.org/pubs/pas/pa589.pdf.

70. In 2004, after reading Kenneth Deffeyes's book on peak oil, *Hubbert's Peak,* I assumed that peak oil was upon us. Obviously, I was wrong. See Bryce, "Bush's Greatest Failure," *Texas Observer,* March 26, 2004. Available: http://www.robert bryce.com/032604.htm.

71. John S. Herold Inc. & Weeden & Co., "Top 10 Energy Investment Themes 2002–2010," February 9, 2007.

72. Interstate Oil and Gas Compact Commission, "Petroleum Professionals," January 2007, 5. Available: http://www.iogcc.state.ok.us/PDFS/2007-Blue-Ribbon -Task-Force-Update.pdf.

73. Author interview, via telephone, June 18, 2007.

74. Joe Carroll, "Chevron Postpones $3 Billion Jack Prospect in Gulf," June 13, 2007. Available: http://www.bloomberg.com/apps/news?pid=20601087&sid =aBohusH_dB2Y&refer=home.

75. Steven Mufson, "U.S. Oil Reserves Get a Big Boost," *Washington Post,* September 6, 2006, D1. Available: http://www.washingtonpost.com/wp-dyn/ content/article/2006/09/05/AR2006090500275.html.

76. That well was drilled about 43 miles south of Morgan City, Louisiana. See Pratt, Priest and Castaneda, *Offshore Pioneers,* back page.

77. Halliburton, "Brown & Root and Kerr-McGee Celebrate 50th Anniversary of First Producing Offshore Oil Well Out-of-Sight-of-Land," November 14, 1997. Available: http://www.halliburton.com/news/archive/1997/bresnws_111497 .jsp.

78. EIA data. Available: http://www.eia.doe.gov/pub/oil_gas/natural_gas/data _publications/crude_oil_natural_gas_reserves/current/pdf/appa.pdf. See table A6.

79. Data from Quest Offshore Resources, Inc., Gulf of Mexico map showing 2006 infrastructure.

80. For more on Statoil, go to its Web page, http://www.statoil.com.

81. John S. Herold Inc. & Weeden & Co., "Top 10 Energy Investment Themes 2002–2010," February 9, 2007.

82. Nationmaster.com says Pakistanis use 1.967 barrels of oil per day per 1,000 people. That equals 0.082 gallons/person/day. Available: http://www.nationmaster .com/graph/ene_oil_con_percap-energy-oil-consumption-per-capita. The same source puts U.S. per capita use at 2.89 gallons/day/person.

83. John Lanchester, "Warmer, Warmer," *London Review of Books*, March 22, 2007. Available: http://www.lrb.co.uk/v29/n06/lanc01_.html.

84. Allison Linn, "Market for Carbon Offsets Raises Questions," May 29, 2007. Available: http://www.msnbc.msn.com/id/18659716.

85. Ed Pilkington, "*An Inconvenient Truth:* Eco-Warrior Al Gore's Bloated Gas and Electricity Bills," *The Guardian*, February 28, 2007. Available: http://environment .guardian.co.uk/energy/story/0,,2022934,00.html.

86. *An Inconvenient Truth*, at approximately 1:30.

87. Liveearth.org. Available: http://www.liveearth.org/?p=138.

88. Liveearth.org. Available: http://liveearthpledge.org/answer_the_call.php.

89. Ari Rabl, Joseph V. Spadaro, Veronika A. Rabl, and Jan F. Kreider, "Costs of Carbon Dioxide Abatement in the United States," *Energy Sustainability 2007*, June 27–30, 2007, 6.

90. Carbon emissions data from EIA, *Annual Energy Outlook 2007*, Figure 8. Available: http://www.eia.doe.gov/oiaf/aeo/excel/figure8_data.xls.

91. Vaclav Smil, "Energy at the Crossroads: Notes for a Presentation at the Global Science Forum Conference on Scientific Challenges for Energy Research," Paris, May 17–18, 2006, 8. Available: http://www.oecd.org/dataoecd/52/25/ 36760950.pdf.

CHAPTER 3

1. Set America Free, "The Hidden Cost of Oil," January 2007. Available: http://www.setamericafree.org/saf_hiddencostofoil010507.pdf.

2. BP, "Statistical Review of World Energy" 2006.

3. Jason S. Grumet, executive director of the National Commission on Energy Policy, prepared statement before the U.S. Senate Committee on Foreign Relations, May 16, 2006. Available: http://lugar.senate.gov/energy/press/pdf/committee _print.pdf.

4. Ethanol across America, "Net Energy Balance of Ethanol Production," Fall 2004, 6. Available: http://www.ethanol.org/pdf/contentmgmt/Issue_Brief_Ethanols _Energy_Balance.pdf.

5. Ali I. al-Naimi, "The Future of U.S.-Saudi Energy Relations," Saudi-U.S. Relations Information Service, May 3, 2006. Available: http://www.saudi-us -relations.org/articles/2006/ioi/060503-naimi-csis.html.

6. Ed Magnuson, "The President Flexes His Muscles," *Time*, October 12, 1981. Available: http://www.time.com/time/magazine/article/0,9171,924894-1,00.html.

7. Leon Hadar, *Sandstorm,* 158.

8. Eugene Gholz and Daryl G. Press, "Energy Alarmism: The Myths That Make Americans Worry about Oil," Cato Institute, April 5, 2007, 15. Available: http://www.cato.org/pubs/pas/pa589.pdf.

9. In the interests of full disclosure, I admit I've made this very same (fallacious) argument. In 2004, I wrote a piece for the *Texas Observer* in which I said that "by depending on Arab states—many of whom fund terrorism—to fill our gas tanks, America dooms itself to fail in the war on terrorism." Available: http://robertbryce.com/092404kerryenergypolicy.htm. Obviously, I have since seen the error of my ways. Mea culpa.

10. Barack Obama, "Updates on Darfur, Immigration, and Gas Prices," April 27, 2006. Available: http://obama.senate.gov/podcast/060427-updates_on_darf/index.php.

11. California Progress Report, "President Clinton: Why I Support Proposition 87 and Why the Oil Companies Are Wrong—The Complete Speech Delivered at UCLA," October 14, 2006. Available: http://www.californiaprogressreport.com/2006/10/first--thank_y.html.

12. Roy Roberson, "Former CIA Director Says Farmers Leading Terrorism Fight," *Southeast Farm Press,* March 8, 2007. Available: http://southeastfarmpress.com/grains/030807-farmers-terrorism.

13. Set America Free, "Set America Free update email," January 29, 2007. Available: www.setamericafree.org/safupdate012907.htm.

14. EIA data. Available: http://tonto.eia.doe.gov/dnav/pet/pet_move_impcus_a2_nus_epc0_im0_mbblpd_a.htm.

15. In 2005, according to the EIA, the U.S. bought crude oil from 41 countries. Available: http://tonto.eia.doe.gov/dnav/pet/pet_move_impcus_a2_nus_epc0_im0_mbbl_a.htm.

16. Zbigniew Brzezinski, "Terrorized by 'War on Terror,'" *Washington Post,* March 25, 2007, B01. Available: http://www.washingtonpost.com/wp-dyn/content/article/2007/03/23/AR2007032301613_pf.html.

17. For more, see the Terror-Free Oil Initiative. Available: http://www.terrorfreeoil.org.

18. http://marketplace.publicradio.org/shows/2007/02/12/PM200702124.html.

19. Sheema Khan, "Don't Be Fearful of Dubai," *Globe and Mail,* March 22, 2006, A19.

20. In 2006, Iraq was the sixth-largest crude supplier to the U.S. according to the EIA. Available: http://www.eia.doe.gov/pub/oil_gas/petroleum/data_publications/company_level_imports/current/import.html.

21. EIA data. Available: http://tonto.eia.doe.gov/dnav/pet/hist/mcrim_nus-nir_2a.htm.

22. Roberson, op. cit. Note that Woolsey is claiming that the Wahhabis are influential in Iran. That's wrong. Wahhabis are Sunni, and Iran is Shiite. According to Stephen Schwartz, a journalist who has written extensively about Wahhabism, the Iranians hate the Wahhabis. For more, see Kathryn Jean Lopez's interview with Schwartz, "The Good and the Bad," National Review Online, November 18, 2002. Available: http://www.nationalreview.com/interrogatory/interrogatory 111802.asp.

23. See the Terror-Free Oil Initiative. Available: http://www.terrorfreeoil.org. By mid–2007, the reference to Saudi Arabia had been removed.

24. The Saudis secretly provided oil to the U.S. military during the Arab oil embargo of 1973. That oil likely helped save the lives of American soldiers in Vietnam. For more on this, see Rachel Bronson, *Thicker than Oil,* 119–120. During the First Iraq War, the Saudis provided much of the fuel needed by the U.S. military. The Saudis also paid much of the cost of the war.

25. Michael Isikoff and Mark Hosenball, "A Legal Counterattack," *Newsweek,* April 16, 2003. Available: http://www.msnbc.msn.com/id/3067906.

26. Bronson, op. cit., 237.

27. Ibid., 244–245.

28. William S. Lind, Col. Keith Nightengale, Capt. John F. Schmitt, Col. Joseph W. Sutton, and Lieut. Col. Gary I. Wilson, "The Changing Face of War: Into the Fourth Generation," *Marine Corps Gazette,* October 1989. Available: http://www .d-n-i.net/fcs/4th_gen_war_gazette.htm.

29. Wilson interview with the author, April 4, 2007, by phone.

30. John F. Burns and Kirk Semple, "U.S. Finds Iraq Insurgency Has Funds to Sustain Itself," *New York Times,* November 26, 2006.

31. http://www.pbs.org/frontlineworld/stories/srilanka/feature2.html.

32. Eleanor Stables, "Terrorism in Spain: Reporters under attack by ETA," *International Herald Tribune,* April 6, 2004. Available: http://www.iht.com/articles/ 2004/04/06/edstables_ed3_.php.

33. The Taliban has, in the past, opposed the cultivation of opium. But over the past few years, America's efforts to eradicate opium cultivation in Afghanistan have resulted in more support for the Taliban by the opium producers. See Ivan Eland, "Another U.S. Escalation in Afghanistan?" Independent Institute, February 26, 2007. Available: http://www.independent.org/newsroom/article .asp?id=1931.

34. Saddam Hussein did pay the families of suicide bombers who attacked Israel. In 2003, reports say that Hussein paid some $260,000 to the families of 26 suicide bombers. See CBS News, "Palestinians Get Saddam Charity Checks," March 13, 2003. Available: http://www.cbsnews.com/stories/2003/03/14/world/ main543981.shtml.

35. Jason Bennetto, "Revealed: MI5 Ruled London Bombers Were Not a Threat," *The Independent,* December 17, 2005. Available: http://news.independent .co.uk/uk/crime/article333649.ece.

36. Gary Kamiya, "Why We Can't Win the 'War on Terror,'" Salon.com, September 15, 2006.

37. Jewish Virtual Library info. Available: http://www.jewishvirtuallibrary.org/ jsource/History/King_David.html.

38. He shared it with Egyptian president Anwar al-Sadat. Available: http:// search.nobelprize.org/search/nobel/?q=sadat&i=en&x=0&y=0.

39. Kati Marton's *A Death in Jerusalem* provides a clear-eyed account of Bernadotte's murder as well as a history of the Stern Gang and Irgun, both of which used terror in order to secure a Zionist state in Palestine.

40. Ibid., 1.

41. Nelson Mandela, from *Long Walk to Freedom: The Autobiography of Nelson Mandela.* London; Little, Brown Book Grooup, 1995: 166.

42. For more, see Robb's blog, Global Guerillas. For his discussion on financing terrorism, see: http://globalguerrillas.typepad.com/globalguerrillas/2004/ 04/global_guerrill.html.

43. Chris Abbott, Paul Rogers, and John Sloboda, "Global Responses to Global Threats: Sustainable Security for the 21st Century," Oxford Research Group, briefing paper, June 2006, 4.

44. U.S. State Department, "State Sponsors of Terrorism," undated. Available: http://www.state.gov/s/ct/c14151.htm.

45. For ethanol, see Marianne Lavelle and Bret Schulte, "Is Ethanol the Answer?" *U.S. News & World Report,* February 12, 2007, 34. Available: http:// www.usnews.com/usnews/news/articles/070204/12ethanol.htm. For biodiesel, see http://www.card.iastate.edu/iowa_ag_review/spring_07/article4.aspx.

46. The Astrodome's volume is 69 million cubic feet. One gallon = 0.1336 cubic feet. Thus, the Astrodome's volume = 516 million gallons.

47. American Airlines, "AMR Corporation Reports a Fourth Quarter Loss of $604 Million, as Compared to a $387 Million Loss in 2004," January 18, 2006. Available: http://www.aa.com/content/amrcorp/pressReleases/2006_01/18_4Q .jhtml.

48. EIA, "Thermal Conversion Factors," Annual Energy Review 2005, 357. Available: http://www.eia.doe.gov/emeu/aer/pdf/pages/sec13_1.pdf.

49. Matthew L. Wald, "Is Ethanol for the Long Haul?" *Scientific American,* January 2007, 46. Wald says the Btu content of ethanol per gallon is about 80,000 Btus, while unleaded regular gasoline contains about 119,000 Btus. Other calculations put the ethanol heat content lower and gasoline content higher. For instance, the Oak Ridge National Laboratory's Web site, the Bioenergy Feedstock

Information Network, puts ethanol's heat content at 75,700 Btus/gallon and gasoline at 125,000 Btus. Available: http://bioenergy.ornl.gov/papers/misc/energy_conv.html.

50. *Houston Business Journal,* "Motiva Sets Major Port Arthur Expansion," November 16, 2006. Available: http://www.bizjournals.com/houston/stories/2006/11/13/daily41.html?from_rss=1.

51. According to the Renewable Fuels Association, there were 110 operating ethanol plants in the U.S. in January 2007. Available: http://www.ethanolrfa.org/industry/statistics.

52. EIA data on prices. Available: http://tonto.eia.doe.gov/dnav/pet/hist/mg_tt_usw.htm. Consumption data: http://tonto.eia.doe.gov/dnav/pet/hist/wgfupus2w.htm.

53. Credit to Robert Rapier of the R-Squared Energy Blog for pointing this out. Available: http://i-r-squared.blogspot.com. See his original article on this: http://i-r-squared.blogspot.com/2007/05/mythical-ethanol-threat.html#links. Also note that the 5 billion gallons of ethanol is equivalent to only about 3 billion gallons of gasoline. Thus, the actual displacement of gasoline would be only about 11.5 percent of the increased gasoline consumption that occurred between 1996 and 2007.

54. The White House, "Twenty in Ten: Strengthening America's Energy Security," undated. Available: http://www.whitehouse.gov/stateoftheunion/2007/initiatives/energy.html.

55. Robert Bryce, "The Great Corn Con," *Slate,* June 26, 2007. Available: http://www.slate.com/id/2169124.

56. EIA data. Available: http://www.eia.doe.gov/neic/quickfacts/quickoil.html.

57. Ibid.

58. Wald, op. cit.

59. This assumes, of course, that all of the new capacity is ethanol. There will be increasing amounts of biodiesel, which has a higher heat content than ethanol. Thus, this is a ballpark number. For import data, see the EIA Web site. Available: http://www.eia.doe.gov/neic/quickfacts/quickoil.html.

60. Government Accountability Office, "Tax Policy: Effects of the Alcohol Fuels Tax Incentives," March 1997, GAO/GGD-97-41, 25.

61. The South Africans have had success with coal-to-liquids processes that make a viable jet fuel. And there have been other small-scale efforts to turn natural gas into a fuel that has been successfully blended with jet fuel. But no manufacturers have been able to create an artificial jet fuel made from plants.

62. EIA, *Annual Energy Outlook 2007,* 57.

63. This estimate of 60 gallons per acre comes from Robert Rapier, "Biodiesel: King of Alternative Fuels," R-Squared Energy Blog, March 27, 2006. Available:

http://i-r-squared.blogspot.com/2006/03/biodiesel-king-of-alternative
-fuels.html. However, other estimates put the soybean biodiesel output significantly lower, at about 52 gallons per acre. See Dennis Avery, "Biofuels, Food or
Wildlife? The Massive Land Costs of U.S. Ethanol," Competitive Enterprise Institute, September 21, 2006, 6. Available: http://www.cei.org/gencon/025,05532
.cfm. At 52 gallons per acre, land needs for biodiesel production would equal 826
million acres.

64. Current cropland totals: about 440 million acres. See: Avery, op. cit.

65. Unlike ethanol, biodiesel is close to oil-based fuel in terms of heat content.
Therefore, volumetric comparison is done on an equal basis. For soybean production figures, see U.S. Department of Agriculture data. Available: http://
www.ers.usda.gov/News/soybeancoverage.htm. Also, 4.8 billion gallons is equal
to 313,000 barrels of oil equivalent per day.

66. Matthias Fawer, "Biofuels—Transporting Us to a Fossil-Free Future?" Bank
Sarasin, July 2006, 12.

67. Hosein Shapouri, James A. Duffield, and Michael Wang, "The Energy Balance of Corn Ethanol: An Update," U.S. Department of Agriculture, July 2002, 9.
Available: http://www.transportation.anl.gov/pdfs/AF/265.pdf.

68. U.S. Department of Agriculture, "Feed Grains Database: Yearbook Tables,"
undated. Available: http://www.ers.usda.gov/data/feedgrains/StandardReports/
YBtable4.htm.

69. Multiply 28.3B × 0.66 = 18.67B gallons. Divide 18.67B gallons by 42 gallons and then by 365 days per year = 1.2M bbls.

70. Again, the U.S. uses about 21 million barrels of oil per day. EIA data. Available: http://www.eia.doe.gov/neic/quickfacts/quickoil.html.

71. Fawer, op. cit., 14.

72. Ibid., 13.

73. Lavelle and Schulte, op. cit., 34.

74. Ibid., 36.

75. Ibid., 34, 39.

76. The calculation is 140 billion/42 gals per bbl/365 days = 9.1 mil bbls/day.
This calculation assumes total 2006 biofuels production of 5.25 billion gallons.
It also assumes that those biofuels would be directly energy-equivalent to oil,
which is not the case. Ethanol contains just two-thirds of the heat energy of
gasoline.

77. EIA, *Annual Energy Outlook 2007*, 8.

78. Ibid., 14.

79. Clearly, most of this production would be domestic and could not be
counted strictly against imports. Instead, those 9.1 million barrels would be produced domestically and would simply provide about one-third of America's total
consumption.

80. EIA petroleum product import numbers. Available: http://tonto.eia.doe .gov/dnav/pet/hist/mttimus2a.htm.

81. California Progress Report, op. cit.

82. Peter Baker, "Bush Says U.S. Pullout Would Let Radicals Use Oil as a Weapon," *Washington Post,* November 5, 2006, A06. Available: http://www.washingtonpost .com/wp-dyn/content/article/2006/11/04/AR2006110401025.html.

83. A. F. Alhajji, "The Failure of the Oil Weapon: Consumer Nationalism vs. Producer Symbolism," *Bridges,* Spring/Summer 2004. Available: http://www2 .onu.edu/~aalhajji/ibec385/pdf/OIL%20Weapon.pdf.

84. Jerry Taylor and Peter van Doren, "Time to Lay the 1973 Oil Embargo to Rest," Cato Institute, October 17, 2003. Available: http://www.cato.org/pub _display.php?pub_id=3272.

85. Alhajji, op. cit., 16.

86. For current inventories in the SPR, see: http://www.spr.doe.gov/dir/ dir.html. According to the EIA, U.S. oil imports are now about 12.5 million barrels per day. At that rate of imports, 700 million barrels would last 56 days.

87. For more on this topic, see Ole Gunnar Austvik, "Strategies for Reducing Consuming Countries' Oil Dependency," May 2004. Available: http://www.kaldor .no/energy/hilrap200405-stratdepoil.html. China began filling its SPR in the summer of 2006. Capacity is expected to ultimately reach about 100 million barrels. See: *Energy Tribune,* "SPRs for Profit?" March 2007, 29.

88. Alhajji, op. cit.

89. Newt Gingrich, "Where Do We Go from Here?" American Enterprise Institute, September 11, 2006. Available: http://www.aei.org/publications/pubID .24891,filter.all/pub_detail.asp.

90. Ed Markey, "Markey: Mass V. EPA Fuels Event Shows Bush Admin. Still Doesn't Get Energy-Global Warming Link," April 10, 2007. Available: http:// globalwarming.house.gov/list/press/global_warming/pr_070410.shtml.

91. Fred Singer, "Old Security Myths," *Washington Times,* November 28, 2003. Available: http://www.sepp.org/Archive/NewSEPP/WashTimes-Security%20Myths %20Singer.html.

92. Marc Labonte and Gail Makinen, "Energy Independence: Would It Free the United States from Oil Price Shocks?" Congressional Research Service, updated January 11, 2002.

93. Thomas L. Friedman, "The Geo-Green Alternative," *New York Times*, January 30, 2005.

94. Institute for the Analysis of Global Security, "Fueling Terror," undated. Available: http://www.iags.org/fuelingterror.html.

95. Alan Reynolds, "Alternative-Fuel Nonsense," National Review Online, May 27, 2005. Available: http://www.nationalreview.com/nrof_comment/reynolds 200505270855.asp.

96. Data from the CIA World Fact Book. Available: https://www.cia.gov/library/publications/the-world-factbook/geos/ba.html.

97. CIA data. Available: https://www.cia.gov/cia/publications/factbook/print/ba.html.

98. Bruce Livesey, "The Salafist Movement," *Frontline*, undated. Available: http://www.pbs.org/wgbh/pages/frontline/shows/front/special/sala.html.

99. The Palestinian elections of 2005 also offer a good example. Rather than elect Fatah, the more moderate wing of Palestine's political system, Palestinian voters chose Hamas, the Islamic party.

100. CIA Factbook. Available: https://cia.gov/cia//publications/factbook/geos/eg.html#Govt.

101. Megan K. Stack and Noha el Hennaway, "Egypt Cracks Down on Muslim Brotherhood," *Los Angeles Times*, February 16, 2007.

102. Ibid.

103. For more, see the index. Available: http://www.heritage.org/research/features/index/countries.cfm.

104. Available: http://www.publishwhatyoupay.org and http://www.eitransparency.org.

105. Thomas L. Friedman, "Thinking about Iraq (I)," *New York Times*, January 22, 2003. See: http://utdocuments.blogspot.com/2006/12/tom-friedmans-pre-war-advocacy-for.html.

106. Lisa J. Adams, "Pemex Celebrates 69th Anniversary, but Problems Loom," Associated Press, March 16, 2007.

107. *Energy Tribune*, "Catarell Heads South," September 2006, 4.

108. S. Galina, D. Romo, F. Ruiz, and A. Pérez, "Hot Spots in Mexico's Oil Industry," presented at the 34th International Energy Conference, International Research Center for Energy and Economic Development, Boulder, Colorado, April 2007.

109. Patrick Harrington, "Lopez Obrador Ends Mexico City Blockade after 47 Days," Bloomberg, September 15, 2006. Available: http://www.bloomberg.com/apps/news?pid=20601086&sid=a93lBNPa1oGQ&refer=latin_america.

110. Robert Bryce, "The Dangers of Cheap Oil," *Energy Tribune*, January 2007, 5.

111. EIA data. Available: http://tonto.eia.doe.gov/dnav/pet/hist/wgtimus2W.htm.

112. EIA data for Venezuela. Available: http://tonto.eia.doe.gov/dnav/pet/hist/mgfimusve2M.htm. For the Netherlands: http://tonto.eia.doe.gov/dnav/pet/hist/mgfimusnl2M.htm.

113. Rex Tillerson, "Remarks by Rex Tillerson, Chairman and CEO, Exxon Mobil Corporation, EFR Business Week 2006, Rotterdam, the Netherlands, April 20, 2006." Available: http://www.exxonmobil.com/Corporate/Newsroom/Spchs Intvws/Corp_NR_SpchIntrvw_RWT_200406.asp.

114. EIA data on gasoline imports. Available: http://tonto.eia.doe.gov/dnav/pet/hist/wgtimus24.htm.

115. *Energy Tribune,* "Reliance Focusing on Exports," May 2007, 33.

CHAPTER 4

1. Peter Tertzakian, "The U.S. Senate's Oil Spill," *Forbes,* November 16, 2005. Available: http://www.forbes.com/2005/11/15/energy-oil-exxonmobil-cx_pt_1116energy_tertzakian.html.

2. *The Independent Financial Review,* "Oil Giants Losing Their Grip," April 4, 2007.

3. Juan Forero, "Orinoco Belt in Venezuela Holds the Promise of Great Oil Riches," *International Herald Tribune,* May 31, 2006. Available: http://www.iht.com/articles/2006/05/31/business/oil.php.

4. Steven Bodzin and Jose Enrique Arrioja, "Conoco, Exxon Mobil to Quit Venezuela, Ramirez Says," Bloomberg, June 26, 2007. Available: http://www.bloomberg.com/apps/news?pid=20601086&sid=aezzzCuHIpvQ&refer=news.

5. Torrey Clark and Lucian Kim, "Gazprom Gains BP Gas Field as Putin Tightens Control," Bloomberg, June 22, 2007. Available: http://www.bloomberg.com/apps/news?pid=newsarchive&sid=afwFHAGtof3Y.

6. Nixon, op. cit.

CHAPTER 5

1. Ronnie Dugger, "Oil and Politics," *Atlantic,* September 1969, 76.

2. For a further discussion of the importance of the East Texas field and the Railroad Commission, see Robert Bryce, *Cronies,* 33–34.

3. EIA data. Available: http://tonto.eia.doe.gov/dnav/pet/hist/mcrntus2a.htm.

4. Texas Railroad Commission. Available: http://www.rrc.state.tx.us/history/h03.html.

5. EIA data. Available: http://tonto.eia.doe.gov/dnav/pet/hist/mcrfpus2a.htm.

6. Ibid.

7. Bryce, *Cronies,* op. cit., 94–95.

8. U.S. Representative George H. W. Bush, "Excerpts from a Speech by Congressman George Bush (R-TEX.), before the Beaumont Gulf Coast Engineering Society, Beaumont, Texas, Friday, February 27, 1970." From George Bush Presidential Library, Congressional file, General Senate Campaign Finance—1970, Box 1, OAID—25858.

9. The import quota on foreign oil was ended in April 1973.

CHAPTER 6

1. http://www.freerepublic.com/focus/f-news/1504102/posts.

2. EIA data show imports rose from 811.1 million barrels in 1972 to 1,183.9 million barrels in 1973 and to 1,269 million barrels in 1974. Available: http://tonto.eia.doe.gov/dnav/pet/hist/mcrimus1a.htm.

3. Donald Losman, "Oil Denial: An Empty Fear," Cato Institute, October 13, 2001. Available: http://www.cato.org/dailys/10-13-01.html.

4. http://www.time.com/time/magazine/article/0,9171,908320,00.html. See also, http://portal.grsu.by/portal/LIBRARY/CD1/politics/cabinet/doehist.htm.

5. Vito A. Stagliano, *A Policy of Discontent*, 23–24.

6. Nixon, op. cit.

7. Glenn Frankel, "U.S. Mulled Seizing Oil Fields in '73," *Washington Post*, January 1, 2004, A01. Available: http://www.washingtonpost.com/ac2/wp-dyn?pagename=article&node=&contentId=A46321-2003Dec31&notFound=true.

8. Jeremy M. Sharp, "U.S. Foreign Aid to Israel," Congressional Research Service, April 25, 2007, CRS-18. Available: http://www.fas.org/sgp/crs/mideast/RL33222.pdf.

9. The date of free fuel shipments was confirmed with the Defense Security Cooperation Agency by the author, by e-mail, January 13, 2006. Between 1998 and 2006, the U.S., through two programs—Foreign Military Sales and Foreign Military Financing—allowed the Israeli military to obtain $1.2 billion in fuel courtesy of American taxpayers. In 2006 alone, the U.S. provided the Israelis with $363.3 million in free motor fuel.

10. Klare, op. cit., 44.

11. The CIA engineered the coup against Mossadegh because he had been insisting that Iranians, not the Anglo-Iranian Oil Company (now BP), should be in control of his country's oil. Anglo-Iranian was making tens of millions of dollars in profits every year from the Iranian concession. And although the British promised the Iranians that they were getting fair compensation for their oil, the Iranians believed—rightly—they were being cheated. For more on this topic, see Stephen Kinzer, *All the Shah's Men*, 50.

12. For the best discussion of how U.S. price controls worsened the gasoline supply situation, read Jerry Taylor and Peter Van Doren, "Economic Amnesia: The Case against Oil Price Controls and Windfall Profit Taxes," Cato Institute, January 12, 2006.

13. Jimmy Carter, State of the Union address, January 23, 1980. Available: http://www.thisnation.com/library/sotu/1980jc.html.

14. Klare, op. cit., 46.

15. Ibid., 47, 48.

16. Available: www.gulfinvestigations.net/document449.html.

17. National Security Archive, "United States Embassy in United Kingdom Cable from Charles H. Price II to the Department of State. 'Rumsfeld Mission: December 20 Meeting with Iraqi President Saddam Hussein,'" December 21, 1983. Available: http://www.gwu.edu/~nsarchiv/NSAEBB/NSAEBB82/iraq31.pdf.

18. Bryce, *Cronies*, 128.

19. Patrick E. Tyler, "U.S. Rejects 'Floating Fortress' in Kuwaiti Waters," *Washington Post*, November 29, 1987, A1.

20. Klare, op. cit., 50.

21. Bryce, *Cronies*, op. cit., 162.

22. PBS timeline. Available: http://www.pbs.org/wgbh/pages/frontline/gulf/cron.

23. Gerald Ford, State of the Union address, January 15, 1975. Available: http://www.ford.utexas.edu/LIBRARY/SPEECHES/750028.htm.

24. Jimmy Carter, televised speech on energy policy, April 18, 1977. Available: http://www.pbs.org/wgbh/amex/carter/filmmore/ps_energy.html.

25. EIA data. Available: http://www.eia.doe.gov/emeu/aer/txt/ptb0518.html.

26. Jimmy Carter, State of the Union Address, January 23, 1980. Available: http://www.thisnation.com/library/sotu/1980jc.html.

27. Stagliano, op. cit., 43.

28. Ibid., 42.

29. Milton R. Copulos, "Salvaging the Synthetic Fuels Corporation," Heritage Foundation, April 12, 1985. Available: http://www.heritage.org/Research/Energy andEnvironment/bg423.cfm.

30. Stagliano, op. cit., 42.

31. Cost figures obtained from Jimmy Carter Presidential Library and Museum by the author, via e-mail, May 14, 2007.

32. Jimmy Carter Presidential Library and Museum, "White House Solar Panel Goes on Display at Carter Library," March 27, 2007. Available: http://www.jimmy carterlibrary.org/newsreleases/2007/07-18.pdf.

33. Ronald Reagan, State of the Union address, February 2, 1985. Available: http://www.thisnation.com/library/sotu/1985rr.html.

34. George H. W. Bush, "Gulf Crisis an Opportunity for a 'New World Order,'" *Congressional Quarterly Weekly Report*, September 15, 1990.

35. George H. W. Bush, "Remarks at a Briefing on Energy Policy, February 20, 1991. Available: http://www.presidency.ucsb.edu/ws/index.php?pid=19318.

36. Ibid.

37. Garry Mauro, "Who Has Best Energy Policy? Clinton Seeks U.S. Energy Independence," *Dallas Morning News*, November 1, 1992.

38. Frank G. Zarb, "How to Win the Energy War," *New York Times*, May 23, 2007.

39. Jimmy Carter Presidential Library and Museum, op. cit.

CHAPTER 7

1. The Institute of Petroleum, "Full Text of Dick Cheney's Speech at the IP Autumn Lunch," November 15, 1999.

2. EIA data. Available: http://tonto.eia.doe.gov/dnav/pet/hist/mcrfpus2a.htm.

3. John S. Herold Inc. & Weeden & Co., "Top 10 Energy Investment Themes 2007–2010," February 9, 2007.

4. John Constable, "The Outlook for Energy—A 2030 View," presentation by John Constable, senior adviser, UK Public Affairs, Exxon Mobil, at the Scottish Parliament, Edinburgh, September 15, 2004. Note that since the 2004 report was released, Exxon Mobil has pushed the non-OPEC peak back slightly, to around 2015.

5. For 2006 revenue data, see: http://money.cnn.com/2006/01/30/news/companies/exxon_earns.

6. Exxon Mobil, 2004 Energy Outlook to 2030. Original available: http://www.exxonmobil.co.uk/files/pa/uk/energy_outlook2004_22.pdf.

7. For more on Exxon Mobil's predictions, see their Energy Outlook to 2030. Available: www.exxonmobil.com/corporate/files/corporate/energy_outlook_2006_notes.pdf.

8. Richard S. Chang, "India's $2,500 Car," *New York Times,* June 22, 2007. Available: http://wheels.blogs.nytimes.com/2007/06/22/indias-2500-car.

9. *Oil and Gas Journal,* "Natural Gas 'Rapidly Gaining in Geopolitical Importance,' Study Finds," June 8, 2004.

CHAPTER 8

1. James Woolsey, "World War IV," FrontPageMagazine.com, November 22, 2002. Available: http://www.frontpagemagazine.com/Articles/Read.aspx?GUID ={2A030550-830B-4638-93C6-417E77B4C321}.

2. Author notes from event.

3. Roy Roberson, "Former CIA Director Says Farmers Leading Terrorism Fight," *Southeast Farm Press,* March 8, 2007. Available: http://southeastfarmpress .com/grains/030807-farmers-terrorism.

4. Project for the New American Century, "Statement of Principles," June 3, 1997. Available: http://www.newamericancentury.org/statementofprinciples.htm.

5. Howard Fineman, "Poetic Justice," MSNBC.com, October 20, 2005. Available: http://www.msnbc.msn.com/id/9752588/site/newsweek.

6. Nina Totenberg, "Lewis Libby Sentenced to 30 Months in Jail," National Public Radio, June 5, 2007. Available: http://www.npr.org/templates/story/story .php?storyId=10741249.

7. Right Web data. Available: http://rightweb.irc-online.org/profile/1100.

8. Right Web data. Available: http://rightweb.irc-online.org/profile/1114.

9. http://www.amazon.com/Rumsfeld-Personal-Portrait-Midge-Decter/dp/0060560916.

10. Project for the New American Century, "Letter to the Honorable William J. Clinton," January 26, 1998. Available: http://www.newamericancentury.org/iraq clintonletter.htm. For more on Zoellick, see Robert Bryce, *Pipe Dreams*, 272, 301.

11. Right Web data. Available: http://rightweb.irc-online.org/profile/1315.

12. Right Web data. Available: http://rightweb.irc-online.org/profile/972.

13. Right Web data. Available: http://rightweb.irc-online.org/profile/1397.

14. Nina J. Easton, "The Hawk; James Woolsey Wants Iraq's Saddam Hussein Brought to Justice," *Washington Post*, December 27, 2001, C01. Note that "slam dunk" would go on to become one of the most infamous phrases of the Second Iraq War. In December 2002, in the Oval Office, during a discussion of the plans to invade Iraq, one of Woolsey's successors as CIA director, George Tenet, assured George W. Bush that the CIA had evidence of weapons of mass destruction in Iraq. "It's a slam dunk case!" Tenet told Bush. When Bush pressed the CIA boss, Tenet responded, "Don't worry, it's a slam dunk." In his 2007 book, *At the Center of the Storm*, for which he received a reported $4 million, Tenet said that his use of the phrase was misconstrued. After the U.S. invasion of Iraq in 2003, no weapons of mass destruction were ever found. For Tenet's quote, see Woodward, *Plan of Attack*, 249. The $4-million figure is from Rich Lowry, "George Tenet's Slam-Dunk," National Review Online, May 1, 2007. Available: http://article.nationalreview .com/?q=ZDliNzNmNjIwNmQxMjM2ZjM2MzYyNWFiYzUyZjgxYTA=. Lowry, the editor of the conservative journal, dismissed Tenet's book as a "classic self-serving tell-all." Tenet's book was a *New York Times* best seller.

15. John Roberts, op. cit., 22.

16. Ariel Cohen and Gerald O'Driscoll, "Privatize Iraqi Oil," National Review Online, December 11, 2002. Available: http://www.nationalreview.com/cohen/cohen121102.asp.

17. For Cohen's bio, see the IAGS Web page. Available: http://www.iags.org/staff.htm#gl. PNAC statement of principles available: http://www.newamerican century.org/statementofprinciples.htm.

18. Gary Rosen, "Neocon Loyalist Recants in Light of Iraqi War," *Houston Chronicle*, March 30, 2006. Available: http://www.chron.com/disp/story.mpl/life/books/reviews/3760031.html.

19. After leaving the Pentagon, Wolfowitz became the head of the World Bank but was forced to resign in mid-2007 over alleged impropriety involving his female companion.

20. Joseph Maldonado, "Analyst: Don't Depend on Oil," *York Daily Record*, March 27, 2003.

21. Sarah Baxter, "Just Think . . . If We Didn't Need Oil There Wouldn't Be a Problem," *Sunday Times,* London: September 19, 2004.

22. Set America Free, "An Open Letter to the American People," September 27, 2004.

23. Institute for Advanced Strategic and Political Studies, "A Clean Break: A New Strategy for Securing the Realm." Available: http://www.iasps.org/strat1 .htm.

24. Robert Bryce, "As Green as a Neocon," *Slate,* January 25, 2005. Available: http://www.slate.com/id/2112608.

25. McFarlane was pardoned by George H. W. Bush in 1992.

26. http://www.iags.org/about.htm.

27. Joe Lieberman and Jon Kyl, "The Present Danger," *Washington Post,* July 20, 2004, A17. Available: http://www.washingtonpost.com/wp-dyn/articles/A63067 -2004Jul19.html.

28. For current signers of the Set America Free document, see: http://www .setamericafree.org/openletter.htm. For Daschle's information, see Lavelle and Schulte, op. cit., or http://www.21stcenturyag.org.

29. http://www.setamericafree.org/index.htm.

30. Data from searches performed on LexisNexis, March 29, 2007.

31. R. James Woolsey, "Gentlemen, Start Your Plug-Ins," OpinionJournal.com, January 1, 2007. Available: http://www.opinionjournal.com/editorial/feature .html?id=110009464. Or see Woolsey's testimony before the U.S. Senate Committee on Energy and Natural Resources on March 7, 2006. Available: http://energy.senate.gov/public/index.cfm?FuseAction=Hearings.Testimony &Hearing_ID=1534&Witness_ID=4342.

CHAPTER 9

1. Yale Center for Environmental Law and Policy, op. cit.

2. Greenpeace USA, "Wind Power," undated. Available: http://www.green peace.org/usa/campaigns/global-warming-and-energy/copy-of-wind-power.

3. Ben Elgin, "An Inconvenient Truth," *Business Week,* March 26, 2007. Available: http://www.businessweek.com/magazine/content/07_13/b4027057.htm.

4. Bernie Woodall, "Wind to Make 20 Percent of Power by 2030: Advocates," Reuters, June 5, 2007. Available: http://www.reuters.com/article/scienceNews/ idUSN0420538920070605.

5. Thomas L. Friedman, "Whichever Way the Wind Blows," *New York Times,* December 15, 2006. Available: http://donkeyod.wordpress.com/2006/12/14/ whichever-way-the-wind-blows.

6. EIA, *Annual Energy Outlook 2006,* Figure 64. Excel data available: http:// www.eia.doe.gov/oiaf/archive/aeo06/excel/figure64_data.xls.

7. Nuclear Energy Institute. Available: http://www.nei.org/resourcesandstats/ nuclear_statistics/usnuclearpowerplants.

8. EIA, *Annual Energy Outlook 2006*, Figure 5. Excel data available: http:// www.eia.doe.gov/oiaf/archive/aeo06/excel/figure5_data.xls.

9. According to the EIA, a barrel of crude contains 5.8 million Btus. Available: http://www.eia.doe.gov/kids/energyfacts/science/energy_calculator.html. A barrel of crude weighs about 315 pounds (http://answers.yahoo.com/question/index? qid=20070628151708AAerQnZ or http://www.funtrivia.com/askft/Question58937 .html). For coal heat value, see EIA data, which put a short ton of coal at 20.7 million Btu. 20.7 million/2,000 lb = 10,377 Btu. Available: http://www.eia.doe.gov/ kids/energyfacts/science/energy_calculator.html.

10. Estimate is from the Agricultural and Biological Engineering Department, Penn State University. Available: http://energy.cas.psu.edu/energycontent.html.

11. Associated Press, "Iowa Plant Tests Grass as Fuel," January 15, 2001. Available: http://www.treepower.org/news/iowa.html.

CHAPTER 10

1. Kurt W. Roth and Kurtis McKenney, "Energy Consumption by Consumer Electronics in U.S. Residences," TIAX LLC, January 2007, 16. Available: http:// www.ce.org/pdf/Energy%20Consumption%20by%20CE%20in%20U.S.%20 Residences%20(January%202007).pdf.

2. Ibid., 14.

3. According to the EIA's *Annual Energy Outlook 2007*, the nonhydro renewable sources of electricity generated 91.74 billion kilowatt-hours of power in 2005. It is implausible to assume that generation could have exceeded 147 billion kilowatt-hours by 2006. Excel data for 2005 available: http://www.eia.doe.gov/ oiaf/aeo/excel/figure60_data.xls.

4. Sean Coughlan, "Do Flat-Screen TVs Eat More Energy?" *BBC News Magazine*, December 7, 2006. Available: http://news.bbc.co.uk/2/hi/uk_news/magazine/ 6188940.stm.

5. Rebecca Smith, "New Power Plants Fueled by Coal Are Put on Hold," *Wall Street Journal*, July 25, 2007, A1. Doubling estimates available: http://en.wikipedia .org/wiki/Doubling_time.

6. Rebecca Smith, "U.S. Electricity Demand Is Outpacing New Resources, Report Warns," *Wall Street Journal*, October 16, 2006, A2.

7. Three Mile Island operated two reactors until the accident in 1979; Exelon data. Available: http://www.exeloncorp.com/NR/rdonlyres/122C645B-58D1-48 DB-9B7D-7FACD0ABEA11/959/ThreeMileIslandPlantFactSheetInternet.pdf.

8. Bureau of Transportation Statistics data. Available: http://www.bts.gov/ publications/national_transportation_statistics/html/table_01_32.html.

9. Ibid.

10. Steve Purdy, "2005 NAIAS—Steve Purdy's Trends and Highlights," TheAuto Channel.com, undated. Available: http://www.theautochannel.com/news/2005/01/15/005099.html.

11. Steven Plotkin, "Is Bigger Better?" *Environment,* November 2004.

12. Don Sherman, "Making Modern Horsepower the Old-Fashioned Way," *New York Times,* January 14, 2007. Available: http://www.nytimes.com/2007/01/14/automobiles/14VIPER.html?ex=1174622400&en=831b4a9b3fbb5c73&ei=5070.

13. Steve Purdy, "2007 Detroit Auto Show—Purdy's Best, Worst and Most Surprising!" TheAutoChannel.com, undated. Available: http://www.theautochannel.com/news/2007/01/15/034046.html.

14. http://www.toyota.com/tundra/index.html.

15. Richard Williamson, "Chevrolet Offers 40 Styles of Silverado," Scripps News.com, March 1, 2007. Available: http://www.scrippsnews.com/node/19738.

16. Carsdirect.com. Available: http://www.carsdirect.com/research/specs?cat=1&make=BM&modelid=444&acode=USB70BMC251A0&year=2007.

17. Gina Chon, "To Woo Wealthy, Lexus Attempts Image Makeover," *Wall Street Journal,* March 24–25, 2007, A1.

18. Mike Spector and Joseph B. White, "Horsepower Nation: New Car Models Boast Speed, Size, Power," *Wall Street Journal,* April 5, 2007, B1.

19. In 1983, gasoline consumption was about 287.1 million gallons per day. In 2005, it was 378.4 million gallons. EIA data available: http://tonto.eia.doe.gov/dnav/pet/hist/c100000001A.htm.

20. EIA, *Annual Energy Outlook 2007,* Table D14, 204.

21. Electricity consumption figures are per EIA's *Annual Energy Outlook 2007,* Table D6, 196.

22. Google. Available: http://www.google.com/corporate/execs.html.

23. John Markoff and Saul Hansell, "Hiding in Plain Sight, Google Seeks More Power," *New York Times,* June 14, 2006.

24. A 10,000-square-foot data center uses as much electricity as 1,000 homes. See Robert Bryce, "Power Struggle," *Interactive Week,* December 19, 2000. Two football fields are equal to about 115,000 square feet. Article available: http://www.robertbryce.com/ar_2.

25. Markoff and Hansell, op. cit.

26. Nate Anderson, "US Servers Now Use More Electricity than Color TVs," *Ars Technica,* February 15, 2007. Available: http://arstechnica.com/news.ars/post/20070215-8854.html.

27. Peter W. Huber and Mark P. Mills, *The Bottomless Well,* 36.

28. Stephen Shankland, "Power Could Cost More than Servers, Google Warns," News.com, May 25, 2006. Available: http://news.com.com/2102-1010_3-5988090.html?tag=st.util.print.

29. EIA data show that the world used 7,404 billion kilowatt-hours in 1980. By 2004, that number was 15,441 billion kilowatt-hours. Available: http://www.eia .doe.gov/pub/international/ielf/table62.xls.

CHAPTER 11

1. Matt Bradley, "What Ever Happened to the Paperless Office?" *Christian Science Monitor,* December 12, 2005. Available: http://www.csmonitor.com/2005/ 1212/p13s01-wmgn.html.

2. *Business Week,* "The 'Soft Path' Solution for Hard-Pressed Utilities," July 23, 1984.

3. EIA data available: http://www.eia.doe.gov/emeu/aer/txt/ptb0801.html.

4. Vaclav Smil, "Energy at the Crossroads: Notes for a Presentation at the Global Science Forum Conference on Scientific Challenges for Energy Research," Paris, May 17–18, 2006, 4. Available: http://home.cc.umanitoba.ca/~vsmil/pdf _pubs/oecd.pdf.

5. *New York Times,* "Energy Ideas, Good and Bad," September 11, 2005.

6. EIA, "World Energy Intensity—Total Primary Energy Consumption per Dollar of Gross Domestic Product Using Market Exchange Rates, 1980–2004," July 31, 2006. Available: http://www.eia.doe.gov/pub/international/ielf/table e1g.xls.

7. EIA, *Annual Energy Outlook 2007,* Table D4, 194.

8. National Commission on Energy Policy, "Ending the Energy Stalemate: A Bipartisan Strategy to Meet America's Energy Challenges," December 2004, 7. Available: http://www.energycommission.org. Note that improving the Corporate Average Fuel Economy standard was widely discussed in Congress in 2007.

9. EIA data. Available: http://tonto.eia.doe.gov/dnav/pet/hist/c100000001M .htm.

10. Bureau of Transportation Statistics data. Available: http://www.bts.gov/ publications/national_transportation_statistics/html/table_01_11.html.

11. Bureau of Transportation Statistics data. Available: http://www.bts.gov/ publications/national_transportation_statistics/html/table_01_25.html.

12. Jerry Adler, "Going Green," *Newsweek,* July 17, 2006.

13. CNNMoney.com, "Hybrid Sales Growth Trailing Off," February 26, 2007. Available: http://money.cnn.com/2007/02/26/autos/hybrid_sales.

14. Peter W. Huber and Mark P. Mills, *The Bottomless Well,* 109.

15. Ibid., 111–112.

16. Vaclav Smil, *Energy at the Crossroads,* 167.

17. Ibid., 332.

18. For more on this topic, see Wikipedia. Available: http://en.wikipedia.org/ wiki/Jevons_paradox.

19. Smil, op. cit., 335.

20. Labonte and Makinen, op. cit.

CHAPTER 12

1. Schmidt interview with the author, March 9, 2007, via telephone.

2. California Progress Report, "President Clinton: Why I Support Proposition 87 and Why the Oil Companies Are Wrong—The Complete Speech Delivered at UCLA," October 14, 2006. Available: http://www.californiaprogressreport.com/2006/10/first-_thank_y.html.

3. Shopfloor.org, "In California, a Bad Proposition," November 3, 2006. Available: http://blog.nam.org/archives/2006/11/in_california_a.php.

4. In 2006, Brazil produced 4.2 billion gallons of ethanol. That's equal to 273,972 barrels per day. Multiplied by 0.66 gives 181,000 bpd.

5. EIA data from 2003. Available: http://www.eia.doe.gov/emeu/states/sep _sum/plain_html/sum_use_tot.html.

6. *An Inconvenient Truth,* text at end of film, at approximately 1:32.

7. JohnEdwards.com, "Achieving Energy Independence and Stopping Global Warming through a New Energy Economy," undated. Available: http://johnedwards .com/about/issues/energy/new-energy-economy.

8. *Energy Tribune,* "More Ethanol Madness," July 2006, 6.

9. Vinod Khosla, "My Big Biofuels Bet," *Wired,* 14.10. Available: http:// www.wired.com/wired/archive/14.10/ethanol_pr.html. Note that in this essay, Khosla claims that in terms of net energy production, gasoline has "half the efficiency of corn ethanol, producing 0.8 BTUs for every BTU input," a claim that is demonstrably false.

10. For more on the tax credit, see the American Coalition for Ethanol, "Net Energy Balance of Ethanol Production," Fall 2004. Available: http://ethanol.org/ index.php?id=78&parentid=26.

11. Environmental Working Group, "Corn Subsidies by Year, U.S. Total," undated. Available: http://www.ewg.org/farm/progdetail.php?fips=00000&prog code=corn.

12. *Washington Post,* "2006 Budget Proposal: Agency Breakdown," February 7, 2005. Available: http://www.washingtonpost.com/wp-srv/politics/interactives/ budget06/budget06Agencies.html. For U.S. Commerce Department employee numbers, see: http://www.whitehouse.gov/omb/budget/fy2004/commerce.html.

13. Environmental Working Group, "Top Programs in United States, 1995–2005," undated. Available: http://www.ewg.org/farm/region.php?fips=00000#topprogs.

14. *Subsidy Watch,* "Costing America's Proposed Alternative Fuel Standard," February 2007, Issue 9. Available: http://www.globalsubsidies.org/article.php3 ?id_article=17&var_mode=calcul#Costing.

15. Global Subsidies Initiative of the International Institute for Sustainable Development, "Biofuels—At What Cost? Government Support for Ethanol and Biodiesel in the United States," October 2006, 52. Available: http://www.global subsidies.org/IMG/pdf/biofuels_subsidies_us.pdf.

16. On August 6, 2007, the nationwide average "rack price" was $2.32. Current prices available: http://www.ethanolmarket.com/fuelethanol.html.

17. Ibid., 56.

18. In 2008, WIC funding was about $5.4 billion. See Office of Management and Budget, "Department of Agriculture," undated. Available: http://www.white house.gov/omb/budget/fy2008/pdf/budget/agriculture.pdf.

19. Government Accountability Office, "Tax Policy: Effects of the Alcohol Fuels Tax Incentives," March 1997, GAO/GGD-97-41, 25.

20. Kopin Tan, "Power by the Bushel," *Barron's,* April 3, 2006, 28. Estimates in early 2007 put ADM's market share at about 25 percent. See Alexei Barrionuevo, "Springtime for Ethanol," *New York Times,* January 23, 2007.

21. *Energy Tribune,* "ADM's New Ethanol Queen," June 2006, 6.

22. Robert J. Samuelson, "A Full Tank of Hypocrisy," *Washington Post,* May 30, 2007, A13. Available: http://www.washingtonpost.com/wp-dyn/content/article/ 2007/05/29/AR2007052901640.html.

23. Reuters, "Ex-Executive at Chevron to Lead Archer Daniels," April 29, 2006. Available: http://www.nytimes.com/2006/04/29/business/29archer.html ?ex=1303963200&en=bacfe27c91be5006&ei=5090&partner=rssuserland&emc =rss.

24. Gregory Palast, "Bayer, ADM, Roche: How a Few Little Piggies Tried to Rig the Market," *Observer,* October 25, 1998. Available: http://www.cbgnetwork.org/ 1219.html.

25. U.S. Department of Justice, "Archer Daniels Midland Co. to Plead Guilty and Pay $100 Million for Role in Two International Price-Fixing Conspiracies," October 15, 1996. Available: http://www.usdoj.gov/opa/pr/1996/Oct96/508at.htm.

26. Online NewsHour, "ADM: Who's Next?" October 15, 1996. Available: http://www.pbs.org/newshour/bb/business/october96/adm_10-15.html.

27. Herbert G. McCann, "Last Defendant in Corn Syrup Case Settles," Associated Press, July 28, 2004.

28. James B. Lieber, *Rats in the Grain,* 326–327.

29. Ibid., 35.

30. Ibid., front matter.

31. Ibid., 38.

32. Ibid., online version, 337. Available: http://www.amazon.com/gp/reader/ 1568582188/ref=sib_dp_pt/002–6565086–1434446#.

33. Kurt Eichenwald, "Archer Daniels Settles Price-Fixing Case," *New York Times,* June 19, 2004.

34. Dan Carney, "Dwayne's World," *Mother Jones*, July/August 1995. Available: http://www.motherjones.com/news/special_reports/1995/07/carney.html.

35. Center for Responsive Politics, "Top All-Time Donor Profiles," undated. Available: http://www.opensecrets.org/orgs/list.asp?order=A.

36. Center for Responsive Politics, "Archer Daniels Midland," undated. Available: http://www.opensecrets.org/orgs/recips.asp?ID=D000000132&Type=P&Cycle=A.

37. Carney, op. cit.

38. Opensecrets.org. Available: http://www.opensecrets.org/orgs/topindivs.asp?ID=D000000132&Display=SC&ContribID=U0000000014.

39. Opensecrets.org. Available: http://www.opensecrets.org/2000elect/other/bush/inaugural.asp.

40. Opensecrets.org. Available: http://www.opensecrets.org/softmoney/softtop.asp?txtCycle=2002&txtSort=amnt.

41. Opensecrets.org. Available: http://www.opensecrets.org/orgs/topindivs.asp?ID=D000000132&Display=SS.

42. James Bovard, "Archer Daniels Midland: A Case Study in Corporate Welfare," Cato Institute, September 26, 1995. Available: http://www.cato.org/pubs/pas/pa-241.html.

43. Ibid.

44. Ben White, "ADM's Largess Preserved Ethanol Break, Study Says," *Washington Post*, June 11, 1998, A21. Available: http://www.washingtonpost.com/wp-srv/politics/campaigns/keyraces98/stories/keycash061198.htm.

45. Shailagh Murray, op. cit.

46. John McCain and John Kyl, "Energy Plan Lacks Juice," *East Valley Tribune*, May 7, 2002. Available: http://mccain.senate.gov/press_office/view_article.cfm?ID=731.

47. John McCain, "Statement of Senator John McCain on the Energy Bill," November 21, 2003. Available: http://mccain.senate.gov/press_office/view_article.cfm?id=274.

48. Ibid.

49. John McCain and John Kyl, "McCain, Kyl Say No To Flawed Energy Bill," June 28, 2005. Available: http://mccain.senate.gov/press_office/view_article.cfm?ID=160.

50. Jon Birger, "McCain's Farm Flip," *Fortune*, October 31, 2006. Available: http://money.cnn.com/magazines/fortune/fortune_archive/2006/11/13/839313/?postversion=2006103111.

51. Shailagh Murray, op. cit.

52. Associated Press, "Clinton denounces memo about her skipping Iowa," May 23, 2007. Available: http://www.usatoday.com/news/politics/2007-05-23-clinton-iowa_N.htm.

53. Iowa Corn, "Ethanol Production Information." Available: http://www.iowacorn.org/ethanol/ethanol_6.html#1.

54. Iowa Corn. Available: http://www.iowacorn.org/ethanol/ethanol_8.html.

55. Iowa Corn. Available: http://www.iowacorn.org/ethanol/ethanol_3a.html.

56. Environmental Working Group, "Total USDA—Subsidies by state, 1995–2005," undated. Available: http://www.ewg.org/farm/progdetail.php?fips=00000&progcode=total&page=states.

57. Environmental Working Group, "Total USDA—Subsidies by year, Iowa," undated. Available: http://www.ewg.org/farm/progdetail.php?fips=19000&progcode=total.

58. Environmental Working Group, "States Receiving corn subsidies, 2005," undated. Available: http://www.ewg.org/farm/progdetail.php?fips=00000&yr=2005&progcode=corn&page=states.

59. According to the U.S. Census Bureau, Iowa has 2.96 million residents. The U.S. has just over 300 million. Available: http://quickfacts.census.gov/qfd/states/19000.html.

60. Frontline, "So You Want to Buy a President?" undated. Available: http://www.pbs.org/wgbh/pages/frontline/president/players/andreas.html.

61. Greenwire, "Society and Politics—Campaign 2000 II: Bradley Defends Ethanol Support," January 10, 2000.

62. RNC Research, "Hillary's Field of Schemes," January 29, 2007. Available: http://www.gop.com/media/PDFs/012907Research.pdf.

63. Ken Silverstein, "Barack Obama Inc.," *Harper's*, November 2006, 40.

64. Shailagh Murray, op. cit.

65. Schmidt interview with the author, March 9, 2007, via telephone.

66. Senator Tom Harkin, "Harkin, Obama, Lugar Introduce Legislation to Increase Availability and Use of Renewable Fuels, Decrease U.S. Dependence on Foreign Oil," January 7, 2007. Available: http://harkin.senate.gov/news.cfm?id=267231.

67. Roberson, op. cit.

68. Lester R. Brown, "Starving the People to Feed the Cars," *Washington Post*, September 10, 2006, B3. Available: http://www.washingtonpost.com/wp-dyn/content/article/2006/09/08/AR2006090801596_pf.html. For more on Brown's group, see http://www.earth-policy.org.

69. Dennis Avery, "Biofuels, Food or Wildlife? The Massive Land Costs of U.S. Ethanol," Competitive Enterprise Institute, September 21, 2006, 6. Available: http://www.cei.org/gencon/025,05532.cfm.

70. Fidel Castro, "Reflections of President Fidel Castro," *Granma*, March 29, 2007. Available: http://www.granma.cu/INGLES/2007/marzo/juev29/14reflex.html.

71. Paulo Sotero and Edward Alden, "Building a Biofuels Alliance," *Washington Post*, March 8, 2007, A23. Available: http://www.washingtonpost.com/wp-dyn/content/article/2007/03/07/AR2007030702046.html.

72. Simla Tokgoz, Amani Elobeid, Jacinto Fabiosa, Dermot J. Hayes, Bruce A. Babcock, Tun-Hsiang (Edward) Yu, Fengxia Dong, Chad E. Hart, and John C. Beghin, "Emerging Biofuels: Outlook of Effects on U.S. Grain, Oilseed, and Livestock Markets," Iowa State University, May 2007, 25. Available: http://www.card .iastate.edu/publications/DBS/PDFFiles/07sr101.pdf.

73. Ibid., 26.

74. American Coalition for Ethanol, op. cit., 11.

75. Shapouri et al., op. cit., 2.

76. Also known as distiller's dried grain.

77. Robert Rapier, "Grain-Derived Ethanol: The Emperor's New Clothes," R-Squared Energy Blog, March 23, 2006. Available: http://i-r-squared.blogspot .com/2006/03/grain-derived-ethanol-emperors-new.html.

78. William K. Jaeger, Robin Cross, and Thorsten M. Egelkraut, "Biofuel Potential in Oregon: Background and Evaluation of Options," Oregon State University, January 29. 2007, 11. Available: http://arec.oregonstate.edu/faculty2/Biofuels %20in%20Oregon%20Jan%2030.pdf.

79. Ibid., ii.

80. Robert Bryce, "Corn Dog," Slate.com, July 19, 2005. Available: http://www .slate.com/id/2122961. See also, David Pimentel and Tad W. Patzek, "Ethanol Production Using Corn, Switchgrass, and Wood; Biodiesel Production Using Soybean and Sunflower," *Natural Resources Research,* March 2005, 65. Available: http://petroleum.berkeley.edu/papers/Biofuels/NRRethanol.2005.pdf.

81. Robert Rapier, "Key Questions on Energy Options," R-Squared Energy Blog, January 21, 2007. Available: http://i-r-squared.blogspot.com/2007/01/ key-questions-on-energy-choices.html#links.

82. Dana Visalli, "Getting a Decent Return on Your Energy Investment," *Energy Bulletin,* April 11, 2006; estimates put the energy profitability of crude oil production at about 10 Btus per every 1 Btu invested. Available: http://www.energy bulletin.net/14745.html. Or see Jan F. Kreider and Peter S. Curtiss, "Comprehensive Evaluation of Impacts from Potential, Future Automotive Fuel Replacements," *Proceedings of Energy Sustainability 2007,* June 27–30, 2007, 11. Their study puts gasoline returns at about 20 Btus per every 1 Btu invested.

83. Government Accountability Office, "Tax Policy: Effects of the Alcohol Fuels Tax Incentives," March 1997, GAO/GGD-97-41, 6.

84. Ibid., 17.

85. EIA, "Emissions of Greenhouse Gases in the United States 2005," November 2006. Available: http://www.eia.doe.gov/oiaf/1605/ggrpt/stopics.html#ethanol.

86. Brent D. Yacobucci, "Fuel Ethanol: Background and Public Policy Issues," Congressional Research Service, October 19, 2006, 22. Available: http://www .ncseonline.org/NLE/CRSreports/06Nov/RL33290.pdf.

87. Alexander E. Farrell, Richard J. Plevin, Brian T. Turner, Andrew D. Jones, Michael O'Hare, and Daniel M. Kammen, "Ethanol Can Contribute to Energy and Environmental Goals," *Science,* January 27, 2006. Abstract available: http://www.sciencemag.org/cgi/content/abstract/311/5760/506.

88. Jerry Taylor and Peter Van Doren, "Ethanol Makes Gasoline Costlier, Dirtier," Cato Institute, January 29, 2007. Available: http://www.cato.org/pub_display.php?pub_id=7308.

89. Kreider and Curtiss, op. cit., 12.

90. Yacobucci, op. cit.

91. The White House, "President Bush and President Lula of Brazil Discuss Biofuel Technology," March 9, 2007. Available: http://www.whitehouse.gov/news/releases/2007/03/20070309-4.html.

92. Peter Baker and Bill Brubaker, "Bush Hails International Ethanol Production," *Washington Post,* March 9, 2007. Available: http://www.washingtonpost.com/wp-dyn/content/article/2007/03/09/AR2007030900767.html.

93. Lauren Etter and Joel Millman, "Ethanol Tariff Loophole Sparks a Boom in Caribbean," *Wall Street Journal,* March 9, 2007, A1.

94. *The Economist,* "Fuel for Friendship," March 3, 2007, 44.

95. Monte Reel, "U.S. Seeks Partnership with Brazil on Ethanol," *Washington Post,* February 8, 2007, A14. Available: http://www.washingtonpost.com/wp-dyn/content/article/2007/02/07/AR2007020702316.html.

96. Edward G. Rendell, "An American Energy Harvest Plan: Jobs, Prosperity, Independence," December 1, 2005. Available: http://www.governor.state.pa.us/governor/cwp/view.asp?a=3&q=444223.

97. Tom Daschle and Vinod Khosla, "Miles Per Cob," *New York Times,* May 8, 2006. Available: http://www.nytimes.com/2006/05/08/opinion/08daschle.html?ex=1304740800&en=8193adacb1a73c25&ei=5090&partner=rssuserland&emc=rss.

98. Marcus Renato Xavier, "The Brazilian Sugarcane Ethanol Experience," Competitive Enterprise Institute, February 15, 2007, 3, 7. Available: http://www.cei.org/gencon/025,05774.cfm.

99. Ibid.

100. Ibid., 3.

101. Ibid. Population figures are from CIA. Available: https://www.cia.gov/cia/publications/factbook/rankorder/2119rank.html.

102. Cambridge Energy Research Associates, "Gasoline Prices, Regulations and Demographics Transforming America's 'Love Affair with the Automobile,'" November 30, 2006. Available: http://www2.cera.com/gasoline/press.

103. Robert Rapier, "Report: Brazilian Ethanol Is Sustainable," R-Squared Energy Blog, October 11, 2006. Available: http://i-r-squared.blogspot.com/2006/

10/report-brazilian-ethanol-is.html#links. See also Luthero Winter Moreira, "Petrobras and Biofuels," Petrobras presentation, May 2007, which says the energy yield on sugarcane is 8.3. Available: http://www2.petrobras.com.br/ri/pdf/Petrobras_Biofuels_May2007.pdf.

104. Food and Agriculture Organization of the United Nations, "Major Food and Agricultural Commodities and Producers," undated. Available: http://www.fao.org/es/ess/top/commodity.html?lang=en&item=156&year=2005.

105. Petrobras data. Available: http://www2.petrobras.com.br/portal/frame.asp?pagina=/ri/ing/Noticias/noticias/Not_etanol.asp.

106. David Luhnow and Geraldo Samor, "As Brazil Fills Up on Ethanol, It Weans Off Energy Imports," *Wall Street Journal*, January 16, 2006. Available: http://yaleglobal.yale.edu/display.article?id=6817.

107. Tom Phillips, "Brazi's Ethanol Slaves: 200,000 Migrant Sugar Cutters Who Prop Up Renewable Energy Boom," *The Guardian*, March 9, 2007. Available: http://www.guardian.co.uk/brazil/story/0,,2029962,00.html.

108. Ibid.

109. Vivian Sequera, "Brazil Raid Frees Ethanol Plant Slaves," Associated Press, July 3, 2007.

110. Monte Reel, "U.S. Seeks Partnership with Brazil on Ethanol," *Washington Post*, February 8, 2007, A14. Available: http://www.washingtonpost.com/wp-dyn/content/article/2007/02/07/AR2007020702316.html.

111. EIA data. Available: http://tonto.eia.doe.gov/dnav/pet/hist/mcrimusbr2m.htm.

112. EIA data. Available: http://tonto.eia.doe.gov/dnav/pet/pet_move_impcus_a2_nus_eppr_im0_mbblpd_a.htm. For petroleum coke, see: http://tonto.eia.doe.gov/dnav/pet/pet_move_impcus_a2_nus_EPPCM_im0_mbblpd_m.htm.

113. EIA data. Available: http://tonto.eia.doe.gov/dnav/pet/hist/mttimusbr2m.htm.

114. Marcus Renato Xavier, "The Brazilian Sugarcane Ethanol Experience," Competitive Enterprise Institute, February 15, 2007, 11. Available: http://www.cei.org/gencon/025,05774.cfm.

115. Ibid.

116. Petrobras data. Available: http://www2.petrobras.com.br/Petrobras/ingles/plataforma/pla_bacia_campos.htm.

117. The Handbook of Texas Online. Available: http://www.tsha.utexas.edu/handbook/online/articles/EE/doe1.html.

118. Petrobras, "National Production of Crude Oil and NGL," undated. In 1997, crude and NGL production was 869,308 barrels per day. Available: http://www2.petrobras.com.br/portal/frame_ri.asp?pagina=/ri/ing/index.asp&lang=en&area=ri. In early 2007, that production was 1,913,500 barrels per day. Available:

http://www2.petrobras.com.br/ri/ing/DestaquesOperacionais/Exploracao
Producao/pdf/ProducaoOleoGas_E&P_2007_Ing.pdf.

119. Xavier, op. cit.

120. Judy Maksoud, "GoM to See $7 Billion in Deepwater Drilling: Petrobras
Ups Domestic Exploration Spending," *Offshore*, May 1, 2007.

121. Dr. Gerhard P. Metschies, "International Fuel Prices 2005," Deutsche Ges-
sellschaft für Technische Zusammenarbeit (GTZ) GmbH, commissioned by
German Federal Ministry for Economic Cooperation and Development, 26.
Available: http://www.gtz.de/de/dokumente/en_International_Fuel_Prices_2005
.pdf.

122. May 2007 gasoline prices obtained from a Brazilian fuel prices Web site:
http://preco.buscape.com.br/combustivel/bp_combustivel2.asp?comb=1&id_cat
eg=6151&est=2. Prices in São Paulo for "gasolina" (approximately 75 percent
gasoline and 25 percent ethanol) ranged from $1.09 to $1.38 per liter. Prices for
pure ethanol were lower, ranging from $0.59 to $0.79 per liter. U.S. gasoline price
is from May 7, 2007, EIA data. Available: http://tonto.eia.doe.gov/dnav/pet/hist/
mg_tt_usw.htm.

123. CIA World Factbook data. Available: https://www.cia.gov/cia/publications/
factbook/rankorder/2174rank.html.

124. CIA World Factbook data on population. Available: https://www.cia.gov/
cia/publications/factbook/rankorder/2119rank.html. The 2.1 million barrels per
day divided by 188 million = 0.011 bbls/day. Multiply by 42 gals/bbl = 0.469 gals/
day/capita.

125. http://www.eia.doe.gov/neic/quickfacts/quickoil.html.

126. *Petrobras Magazine*, No. 122, "Ethanol: From Brazil to the World," undated. Available: http://www2.petrobras.com.br/portal/frame.asp?pagina=/
ri/ing/Noticias/noticias/Not_etanol.asp.

127. Maksoud, op. cit.

128. Amory B. Lovins, E. Kyle Datta, Odd-Even Bustnes, Jonathan G. Koomey,
and Nathan J. Glasgow, *Winning the Energy Endgame*, executive summary, 3.
Available: http://www.ethanol-gec.org/information/briefing/5.pdf.

129. Amory B. Lovins, testimony before the U.S. Senate Committee on Energy
and Natural Resources, March 7, 2006. Available: http://energy.senate.gov/public/
index.cfm?FuseAction=Hearings.Testimony&Hearing_ID=1534&Witness
_ID=4345.

130. Richard G. Lugar and R. James Woolsey, "The New Petroleum," *Foreign
Affairs*, January/February 1999. Available: http://www.foreignaffairs.org/
19990101faessay954/richard-g-lugar-r-james-woolsey/the-new-petroleum.html.

131. Set America Free, "A Blueprint for U.S. Energy Security," 5. Available:
http://www.setamericafree.org/blueprint.pdf.

132. David Roberts, "Al Gore, Movie Star, Talks of His Latest Role," *Grist*, May 24, 2006. Available: http://www.msnbc.msn.com/id/12743273.

133. California Progress Report, "President Clinton," op. cit.

134. John Kerry, undated. Available: www.thedemocraticdaily.com/Kerry _Energy_Plan_Fact_Sheet.doc.

135. George W. Bush, State of the Union address, January 31, 2006. Available: http://www.whitehouse.gov/news/releases/2006/01/20060131-10.html.

136. George W. Bush, State of the Union address, January 23, 2007. Available: http://www.whitehouse.gov/news/releases/2007/01/20070123-2.html.

137. The White House, "President Bush Participates in Panel on Cellulosic Ethanol," February 22, 2007. Available: http://www.whitehouse.gov/news/ releases/2007/02/20070222-5.html.

138. The White House, "Fact Sheet: Harnessing the Power of Technology for a Secure Energy Future," February 22, 2007. Available: http://www.whitehouse.gov/ news/releases/2007/02/20070222-2.html.

139. *New York Times,* "The Warming Challenge," May 5, 2007. Available: http://www.nytimes.com/2007/05/05/opinion/05sat1.html.

140. *New York Times,* "Testing Time on Energy," September 3, 2007. Available: http://www.nytimes.com/2007/09/03/opinion/03mon1.html.

141. http://www.alabev.com/history.htm.

142. Thanks to Dr. Donald Stedman for this concept.

143. Michael Oneal, "Scientists Seek Cheap, Plentiful Energy Alternatives," *Chicago Tribune,* October 13, 2006. Available: http://www.chicagotribune.com/ business/chi-0610130128oct13,0,7209827.story?coll=chi-business-hed.

144. Robert Rapier, "Cellulosic Ethanol vs. Biomass Gasification," R-Squared Energy Blog, October 23, 2006. Available: http://i-r-squared.blogspot.com/ 2006/10/cellulosic-ethanol-vs-biomass.html#links.

145. Bill Hord, "The Future Is Not Now for Biomass Ethanol Industry," *Omaha World-Herald,* March 27, 2007. Available: http://www.omaha.com/index.php ?u_page=1208&u_sid=2354471.

146. Ibid.

147. Robert Rapier, "The Logistics Problem of Cellulosic Ethanol," R-Squared Energy Blog, March 28, 2007. Available: http://i-r-squared.blogspot.com/2007/ 03/logistics-problem-of-cellulosic-ethanol.html#links.

148. Available: http://i-r-squared.blogspot.com.

149. Rapier, March 28, 2007, op. cit.

150. Bryce, op. cit.

151. Oneal, op. cit.

152. John R. Benemann, Don C. Augenstein, Don J. Wilhelm, and Dale R. Simbeck, "Ethanol from Lignocellulosic Biomass—A Techno-Economic Assessment," presented at IEA Bioenergy Workshop, Vancouver, August 29–30, 2006. Abstract available: http://www.theoildrum.com/story/2006/8/15/13634/6716.

153. Confirmed by author with Caruso via e-mail, February 26, 2007.

154. Jaeger et al., op. cit.

155. Tokgoz et al., op. cit., abstract page.

156. http://www.enchantedlearning.com/usa/states/area.shtml.

157. John Deutch, "Biomass Movement," *Wall Street Journal,* May 10, 2006, A18. Available: http://downloads.heartland.org/19037.pdf.

158. At 1 million barrels per 39,000 square miles, 10.5 million barrels would require 409,500 square miles. Texas contains 268,581 square miles.

159. Government Accountability Office, "Tax Policy: Effects of the Alcohol Fuels Tax Incentives," March 1997, GAO/GGD-97-41, 24.

160. Hord, op. cit.

161. EIA, "Status and Impact of State MTBE Ban," March 27, 2003. Available: http://www.eia.doe.gov/oiaf/servicerpt/mtbeban/index.html.

162. Cattlenetwork.com, "Ethanol Outlook: Summer Production Expected to Average 399,000 Bbl/d," April 10, 2007. Available: http://www.cattlenetwork.com/content.asp?contentid=120507.

163. The metric for volatility is Reid vapor pressure, or RVP. Ethanol, when blended with gasoline, raises the RVP of the blended gasoline.

164. These volume losses were confirmed with refinery personnel at both Exxon Mobil and Shell.

165. Government Accountability Office, "Gasoline Markets: Special Gasoline Blends Reduce Emissions and Improve Air Quality, but Complicate Supply and Contribute to Higher Prices," June 2005, GAO-05-421, 30.

166. *Consumer Reports,* "The Ethanol Myth," October 2006. Available: http://www.consumerreports.org/cro/cars/new-cars/ethanol-10-06/overview/1006_ethanol_ov1_1.htm.

167. Government Accountability Office, "Crude Oil: Uncertainty about Future Oil Supply Makes It Important to Develop a Strategy for Addressing a Peak and Decline in Oil Production," February 2007, GAO-07-283, highlights page. Available: http://www.gao.gov/new.items/d07283.pdf.

168. Bob Dinneen, "Biofuels Are Our Future, but We Need Much More than Just Corn," *Wall Street Journal,* May 23, 2007, A15.

169. http://www.epa.gov/epahome/aboutepa.htm.

170. Environmental Protection Agency, "Bush Administration Establishes Program to Reduce Foreign Oil Dependency, Greenhouse Gases," April 10, 2007. Available: http://yosemite.epa.gov/opa/admpress.nsf/e87e8bc7fd0c11f1852572a000650c05/9f276d4de20fe075852572b9005cb19c!OpenDocument.

171. Environmental Protection Agency, "EPA Finalizes Regulations for a Renewable Fuel Standard (RFS) Program for 2007 and Beyond," April 2007, 3. Available: http://www.epa.gov/otaq/renewablefuels/420f07019.pdf.

172. Author interview with Wood, by phone, April 19, 2007.

173. Author interview with Becker, by phone, April 20, 2007.

174. EPA, "Six Common Air Pollutants," undated. Available: http://www.epa .gov/air/urbanair/nox/effrt.html.

175. EPA, "Health and Environmental Impacts of NO_x," undated. Available: http://www.epa.gov/air/urbanair/nox/hlth.html.

176. EPA, "Ground-Level Ozone," undated. Available: http://www.epa.gov/air/ ozonepollution/basic.html.

177. *U.S. Federal News,* "Sen. Feinstein Comments on EPA's Rejection of California's Application of Oxygenate Waiver," June 2, 2005. See also, *Oil Daily,* "Study Says Ethanol in Gasoline Increases Evaporative Emissions," September 17, 2004.

178. Ibid.

179. The air district covers all or part of four counties. Available: http://www .aqmd.gov/map/MapAQMD1.pdf.

180. *Energy Washington Week,* "California Air Regulators May Want to Control Ethanol in State," July 19, 2006.

181. Mark Z. Jacobson, "Effects of Ethanol (E85) versus Gasoline Vehicles on Cancer and Mortality in the United States," *Environmental Science and Technology,* April 18, 2007. Available: http://www.stanford.edu/group/efmh/jacobson/ es062085v.pdf.

182. Author interview with Stedman, by phone, April 19, 2007.

183. EPA, "EPA Finalizes Regulations," op. cit.

184. U.S. Department of Energy, "Energy Demands on Water Resources," December 2006, 9. Available: http://www.sandia.gov/energy-water/docs/121-Rpt ToCongress-EWwEIAcomments-FINAL.pdf.

185. Ibid., 61–62.

186. The math is straightforward: 1 million Btus / 80,000 Btus per gallon of ethanol = 12.5 gallons of ethanol per MMBtu.

187. EPA, "Exercise II: The Superior Car Wash," undated. Available: http:// www.epa.gov/nps/nps_edu/stopx2.htm.

188. Pimentel and Patzek, op. cit., 66.

189. http://www.nebraskacorn.org/cornmerch/usdareports.htm.

190. Fifteen percent of 885 is 132.75.

191. U.S. Department of Energy, op. cit., 57, 59. Note that the report estimates that oil extraction requires between 5 and 13 gallons of water per barrel of crude, or, at most, 0.3 gallons of water per gallon of crude. Page 59 puts the refining requirements at "about 1 to 2.5 gallons" of water for each gallon of product.

192. Ibid., 62. The report says water consumption in ethanol processing averages 62 gallons per MMBtu. Thus, 12.5 gallons of ethanol = 1 MMBtu. And 62 / 12.5 = 4.96 gallons of water per each gallon of ethanol.

193. One acre-foot of water contains 325,851 gallons. For data on Oklahoma lakes, see the Oklahoma Water Development Board, "Oklahoma Water Facts," undated. Available: http://www.owrb.ok.gov/util/waterfact.php.

194. California uses about 43.1 million acre-feet of water per year. See Samantha Young, "Debate 'Warming' over Calif. Water Supply," *USA Today,* April 7, 2007. Available: http://www.usatoday.com/news/nation/2007-04-07-california water_N.htm. State population in 2005 was 36.1 million. See Census Bureau data. Available: http://quickfacts.census.gov/qfd/states/06000.html.

195. Southern Nevada Water Authority, "Colorado River Facts," undated. Available: http://www.snwa.com/html/wr_colrvr.html. For length data, see U.S. Geological Survey, "Largest Rivers in the United States," May 1990. Available: http://pubs.usgs.gov/of/1987/ofr87-242.

196. Kreider and Curtiss, op. cit., 11.

197. Cornell University, "End Irrigation Subsidies and Reward Conservation, Cornell Water-Resources Study Advises," January 20, 1997. Available: http://www.news.cornell.edu/releases/Jan97/water.hrs.html.

198. Bill Lambrecht, "Ethanol Plants Come with Hidden Cost: Water," *St. Louis Post-Dispatch,* April 15, 2007.

199. Maria Sudekum Fisher, "Ethanol Boom Creates Dilemma for Farmers and Small Towns," Associated Press, March 26, 2007. Available: http://www2.ljworld.com/news/2007/mar/26/ethanol_boom_creates_dilemma_farmers_and_small_tow.

200. *Wichita Eagle,* "Time to Get Serious about Ogallala Aquifer," April 8, 2007. Available: http://blogs.kansas.com/weblog/2007/04/time_to_get_ser.html.

201. Ibid., 62.

202. The White House, "President Bush Participates in Demonstration of Alternative Fuel Vehicles with CEOs of Ford, General Motors and Daimler-Chrysler," March 26, 2007. Available: http://www.whitehouse.gov/news/releases/2007/03/20070326.html.

203. Ford Motor Co. Available: http://www.ford.com/en/innovation/technology/ethanolCapableVehicles/default.htm.

204. General Motors. Available: http://www.gm.com/company/gmability/environment/images/e85/flexfuelposter.pdf.

205. DaimlerChrysler. Available: http://www.daimlerchrysler.com/dccom/0-5-632674-1-632704-1-0-0-635864-0-0-135-631970-0-0-0-0-0-0-0.html.

206. Green Car Congress, "GM: Live Green Go Yellow," January 25, 2006. Available: http://www.greencarcongress.com/2006/01/gm_live_green_g.html.

207. http://www.daimlerchrysler.com/dccom/0-5-632674-1-632704-1-0-0-635864-0-0-135-631970-0-0-0-0-0-0-0.html.

208. Ford Motor Co. Available: http://www.ford.com/en/innovation/technology/ethanolCapableVehicles/default.htm.

209. Paul Rauber, "Decoder: Corn-Fed Cars," *Sierra,* January/February 2007. Available: http://www.sierraclub.org/sierra/200701/decoder.asp.

210. John Kerry, "Our Energy Challenge," June 26, 2006. Available: http://www.truthout.org/cgi-bin/artman/exec/view.cgi/61/20762.

211. Brent Yacobucci, "Ethanol and Biofuels: Agriculture, Infrastructure, and Market Constraints Related to Expanded Production," Congressional Research Service, March 16, 2007. Available: http://fpc.state.gov/documents/organization/82500.pdf.

212. In 2006, there were about 4.5 million FFVs in the U.S. that could run on E85. Assuming that the big Detroit automakers produced another 1 million FFVs in 2007, there would be about 5.5 million vehicles in the U.S. capable of using E85 by 2008. There are about 247 million motor vehicles in the U.S. Thus, about 2.2 percent of all U.S. vehicles would be E85-capable.

213. Ed Wallace, "Ethanol: A Tragedy in 3 Acts," *Business Week,* April 27, 2006. Available: http://www.businessweek.com/autos/content/apr2006/bw20060427_493909.htm?campaign_id=topStories_ssi_5.

214. Amanda Griscom Little, "Corn at the Right Time," *Grist,* February 24, 2006. Available: http://www.grist.org/news/muck/2006/02/24/griscom-little.

215. Rauber, op. cit.

216. Laura Meckler, "Fill Up with Ethanol? One Obstacle Is Big Oil," *Wall Street Journal,* April 2, 2007, A1.

217. *Consumer Reports,* op. cit.

218. Lavelle and Schulte, op. cit., 39.

219. Green Car Congress, op. cit.

220. EIA data. Available: http://www.eia.doe.gov/kids/history/timelines/ethanol.html.

221. MSN Autos, "America's 'Greenest' Vehicles." Available: http://autos.msn.com/advice/article.aspx?contentid=4018862.

222. Micheline Maynard, "Move Over G.M., Toyota Is No. 1," *New York Times,* April 25, 2007. Note that this story says that there were two periods during the 1970s and 1990s when strikes caused GM to briefly lose its top spot.

CHAPTER 13

1. EIA data. Available: http://tonto.eia.doe.gov/dnav/ng/hist/n9050us2a.htm.

2. EIA data. Available: http://tonto.eia.doe.gov/dnav/ng/hist/n9102cn2a.htm.

3. Energy Tribune, "Canada Gas Peak," May 2006, 6.

4. For doubling figures, see EIA data, available: http://tonto.eia.doe.gov/dnav/ng/hist/n9103us2a.htm. The EIA expects LNG to supply 4.5 trillion cubic feet (Tcf) of gas by 2030. For 2030 import data, see EIA, Annual Energy Outlook 2007, Figure 77. The EIA expects total imports to be 5.7 Tcf out of 26.1 Tcf of consumption. Imports spreadsheet available: http://www.eia.doe.gov/oiaf/aeo/excel/figure77_data.xls. For consumption data, see EIA, Annual Energy Outlook 2007, 3.

5. Energy Tribune, "Canada Gas Peak," op. cit.

6. EIA data. Available: http://tonto.eia.doe.gov/dnav/ng/ng_move_impc_s1_a.htm.

7. Several companies are pursuing the CNG technology, including EnerSea and SeaNG.

CHAPTER 14

1. The other owners of the plant are the city of San Antonio's electric utility, CPS Energy, which owns 40 percent, and publicly traded NRG Energy, which owns 44 percent. In September 2007, NRG filed a license application with the Nuclear Regulatory Commission to build two more reactors at the site of the South Texas Project.

2. Mike Clark-Madison, "The Roots of the Left," *Austin Chronicle,* July 4, 2003. Available: http://www.austinchronicle.com/gyrobase/Archive/search?searchType=archives&Search=South%20Texas%20Nuclear%20Project.

3. STP Nuclear Operating Company. Available: http://www.stpnoc.com/STP%20Milestones%20-%20web%20copy.htm.

4. Clark-Madison, op. cit.

5. STP Nuclear Operating Company, "STP Leads U.S. in Production for Third Straight Year," February 15, 2007. Available: http://www.stpnoc.com/PrRel%2007-1B%20Prod%20Records.doc.

6. Figures are from Austin Energy's media relations department. Data received by the author via e-mail, March 22, 2007.

7. All fuel cost figures are from Austin Energy's media relations department. Data received by the author via e-mail, February 20, 2007.

8. This is an estimate. The city of Austin refuses to provide detailed information on its savings. However, gas prices were generally higher in the years following 2001. In 2001, gas for power generation cost $4.61. By 2005, the price had hit $8.48. Data from the EIA, available: http://tonto.eia.doe.gov/dnav/ng/ng_pri_sum_dcu_nus_a.htm.

9. The city's utility, Austin Energy, refuses to disclose the plant's operating costs, saying the information is proprietary. The $0.017 figure is from the Nuclear Energy Institute estimates. NEI data available: http://www.nei.org/resourcesandstats/nuclear_statistics/costs. The NEI estimate is corroborated by data from the California Energy Commission, which estimates natural-gas-fired electricity costs to be at least $0.052 per kilowatt-hour. The agency puts nuclear power at $0.014 to $0.019. Available: http://www.energy.ca.gov/electricity/comparative_costs.html.

10. Generation costs were provided by the city of Austin utility.

11. Austin Energy data, received by author, February 20, 2007.

12. Nuclear Energy Institute data. Available: http://www.nei.org/resourcesand stats/nuclear_statistics/usnuclearpowerplants.

13. Julian Steyn, "Fuel Review: Supply—A Mature Market?" *Nuclear Engineering International,* September 12, 2005, 10. See also EIA data, available: http://www.eia.doe.gov/cneaf/nuclear/umar/umar.html.

14. EIA data. Available: http://www.eia.doe.gov/cneaf/nuclear/umar/table3.html.

15. Ellen Miller, "Soviets, Responsible for Uranium Boom, Now Blamed for Bust," Associated Press, November 30, 1990.

16. *New Mexico Business Journal,* "Money in Mining?" August 1, 1985.

17. EIA data. Available: http://www.eia.doe.gov/cneaf/nuclear/dupr/dupr.html.

18. European Nuclear Society, "Uranium Reserves," undated. Available: http://www.euronuclear.org/info/encyclopedia/u/uranium-reserves.htm.

19. Paul Foy, "Uranium Mining Is Reborn," Associated Press, September 17, 2006.

CHAPTER 15

1. EIA data from *Annual Energy Outlook 2007.*

2. EIA data. Available: http://www.eia.doe.gov/cneaf/coal/quarterly/html/ t18p01p1.html. For import/export data, see EIA. Available: http://www.eia .doe.gov/neic/quickfacts/quickcoal.html. It's also worth noting that although the U.S. is a big crude oil importer, it's also a big *exporter* of oil products. For instance, in August 2007, it was exporting about 1.2 million barrels of petroleum products per day. See EIA data: http://tonto.eia.doe.gov/dnav/pet/pet_move _wkly_dc_NUS-Z00_mbblpd_w.htm.

3. Mark Clayton, "Why Coal-Rich US Is Seeing Record Imports," *Christian Science Monitor,* July 10, 2006. Available: http://www.csmonitor.com/2006/ 0710/p02s01-usec.html.

4. EIA data. Available: http://www.eia.doe.gov/cneaf/coal/quarterly/html/ t7p01p1.html. The U.S. used 1.1 billion tons of coal in 2006. See EIA data. Available: http://www.eia.doe.gov/cneaf/coal/quarterly/html/t28p01p1.html.

5. EIA data. Available: http://www.eia.doe.gov/cneaf/coal/quarterly/html/ t28p01p1.html.

6. EIA data from *Annual Energy Outlook 2007,* Figure 87. Available: http://www.eia.doe.gov/oiaf/aeo/excel/figure87_data.xls.

7. Brian Schweitzer, testimony before Senate Finance Committee, February 27, 2007. Available: http://www.senate.gov/~finance/hearings/testimony/2007 test/022707bstest.pdf.

8. Jim Efstathiou, Jr., "Bush Presses for Coal Liquids to Cut Gasoline Use," Bloomberg, April 12, 2007. Available: http://www.bloomberg.com/apps/news ?pid=20601072&refer=energy&sid=afE2yWvJ6NoM.

9. Barack Obama, "Senators Obama and Bunning Introduce Legislation to Expand Coal Use," June 7, 2006. Available: http://obama.senate.gov/press/060607-senators_obama_and_bunning_introduce_legislation_to_expand_coal_use/index.html.

10. Vito A. Stagliano, *A Policy of Discontent*, 8.

11. Gerald Ford, State of the Union address, January 15, 1975. Available: http://www.ford.utexas.edu/LIBRARY/SPEECHES/750028.htm.

12. Stagliano, op. cit., 9.

13. Edmund L. Andrews, "Lawmakers Push for Big Subsidies for Coal Process," *New York Times*, May 29, 2007.

14. Robert Rapier, "The Cost of Environmental Regulations," R-Squared Energy Blog, June 1, 2007. Available: http://i-r-squared.blogspot.com/2007/06/cost-of-environmental-regulations.html#links.

15. Andrews, op. cit.

16. *Energy Tribune,* "CTL Plants Boom," July 2006, 18.

17. Masayuki Sasanouchi, "Fuel Related Challenges of Environmental Auto Technology," Toyota Motor Corporation, September 27, 2005. PowerPoint presentation by Sasanouchi obtained by the author.

18. EIA, *Annual Energy Outlook 2007,* 96.

19. The EIA expects oil consumption in the U.S. to hit 27 million barrels per day by 2030. See EIA, *Annual Energy Outlook 2007,* 8.

20. It should be noted that China is currently in the midst of a massive CTL expansion. Over the next half decade or so, China expects to have CTL production capacity of 600,000 barrels per day.

CHAPTER 16

1. For square footage, see Margot Adler, "Behind the Ever-Expanding American Dream House," National Public Radio, July 4, 2006. Available: http://www.npr.org/templates/story/story.php?storyId=5525283. For energy use, see EIA data. Available: http://www.eia.doe.gov/emeu/recs/recs2001/enduse2001/enduse2001.html.

2. EIA data: The average residential customer pays about $0.095 per kilowatt-hour. Available: http://www.eia.doe.gov/cneaf/electricity/epa/epat7p4.html.

3. For instance, a 15-year loan at 6 percent on that $7,445 would require payments of about $63 per month. The average monthly savings from the solar panels is about $32.

4. In April 2007, Pfizer was paying a dividend of 4.4 percent. Several closed-end bond funds were paying in excess of 5 percent, including Western Asset 2008 Worldwide Dollar Government Term Trust Inc. Ticker: SBG.

5. Pavel Molchanov, "Alternative Energy Earnings Preview for 1Q07; Adjusting Corn Forecast," Raymond James & Associates, April 10, 2007, 4.

6. Pearce Hammond and Brian Gamble, "Simmons Oil Monthly—Solar Energy Overview," Simmons & Company International, February 16, 2006, 1.

7. Ibid., 3.

CHAPTER 17

1. Al Gore, "My Turn: The Energy Electranet," *Newsweek,* December 18, 2006. Available: http://www.msnbc.msn.com/id/16127831/site/newsweek.

2. Greenpeace, "Wind Power," undated. Available: http://www.greenpeace.org/usa/campaigns/global-warming-and-energy/copy-of-wind-power.

3. NRDC, "Wind Power," undated. Available: http://www.nrdc.org/air/energy/renewables/wind.asp.

4. U.S. Department of Energy, "Wind Powering America," April 2000. Available: http://www.nrel.gov/docs/fy00osti/28133.pdf.

5. American Council on Renewable Energy, *The Outlook on Renewable Energy in America, Volume 2: Joint Summary Report,* March 2007, 15. Available: http://www.acore.org/pdfs/ACORE_Joint_Outlook_Report.pdf.

6. The U.S. had 978 gigawatts of electric capacity installed in 2007. EIA data. Available: http://www.eia.doe.gov/neic/quickfacts/quickelectric.html.

7. National Research Council, "Environmental Impacts of Wind-Energy Projects," executive summary, May 2007, 1. Available: http://www.nap.edu/execsumm_pdf/11935.pdf.

8. ESB National Grid, "Impact of Wind Generation in Ireland on the Operation of Conventional Plant and the Economic Implications," 2004, 36. Available: http://www.eirgrid.com/EirGridPortal/uploads/Publications/Wind%20Impact%20Study%20-%20main%20report.pdf.

9. Ibid.

10. Pete du Pont, "Air Power," *Opinion Journal,* April 25, 2007. Available: http://www.opinionjournal.com/columnists/pdupont/?id=110009980.

11. Available: www.ercot.com/news/press_releases/2007/ERCOT_Response_to_Rep._Barton.html.

12. Tom McGhie, "Wind Turbines 'Not reliable,'" *Financial Mail,* April 15, 2007. Available: http://www.thisismoney.co.uk/news/article.html?in_article_id=419366&in_page_id=2.

13. Hugh Sharman, "Why Wind Power Works for Denmark," *Civil Engineering 158,* May 2005, 69. Available: http://www.thomastelford.com/journals/Document Library/CIEN.158.2.66.pdf.

14. Luke Harding, John Vidal, and Alok Jha, "Report Doubts Future of Wind Power," *The Guardian,* February 26, 2005. Available: http://society.guardian.co.uk/environment/story/0,14124,1425868,00.html.

15. Royal Academy of Engineering, "The Cost of Generating Electricity," 2004, 3–5. Available: http://www.raeng.org.uk/news/publications/list/reports/Cost _Generation_Commentary.pdf.

16. Harding et al., op. cit.

17. Country Guardian. Available: http://www.countryguardian.net/Manifesto .htm.

18. Capewind.org. Available: http://www.capewind.org/article24.htm.

19. Robert F. Kennedy, Jr., "An Ill Wind off Cape Cod," *New York Times,* December 16, 2005. Available: http://www.nytimes.com/2005/12/16/opinion/ 16kennedy.html?ex=1292389200&en=58e5dd67e381fd58&ei=5090&partner =rssuserland&emc=rss.

20. Michael Shellenberger and Ted Norhaus, "Arctic Battle Should Move to Hyannis Port," *San Francisco Chronicle,* December 21, 2005. Available: http://www.sfgate.com/cgi-bin/article.cgi?f=/c/a/2005/12/21/EDGU6 GALTN1.DTL.

21. Robert Elder, "King Ranch Leads Backlash against Wind Farms," *Austin American-Statesman,* March 28, 2007.

22. John Porretto, "A Texas-Size Fight over Wind Power," Associated Press, June 24, 2007. Available: http://www.statesman.com/business/content/business/ stories/technology/06/24/24wind.html.

23. American Council on Renewable Energy, op. cit., 27.

24. Edison Electric Institute, Electricity 101, undated, 32. Available: http://www .eei.org/industry_issues/industry_overview_and_statistics/Electricity_101.pdf.

25. Clifford Krauss, "Where Now, for the Wind?" *New York Times,* June 1, 2007.

26. U.S. Department of Energy, op. cit.

27. EIA, *Annual Energy Outlook 2006,* Figure 64. Available: http://www.eia.doe .gov/oiaf/archive/aeo06/excel/figure64_data.xls.

28. EIA, *Annual Energy Outlook 2007,* 7. Available: www.eia.doe.gov/oiaf/ aeo/index.html.

29. Thomas L. Friedman, *The World Is Flat,* 8.

30. Yale Global Online, "'Wake Up and Face the Flat Earth'—Thomas L. Friedman," April 18, 2005. Available: http://yaleglobal.yale.edu/display.article?id=5581.

31. Thomas L. Friedman, "The Energy Wall," *New York Times,* December 1, 2006, 31.

32. Thomas L. Friedman, "Let's Roll," *New York Times,* January 2, 2002, 15.

33. Thomas L. Friedman, "A Failure to Imagine," *New York Times,* May 19, 2002, 15.

34. Thomas L. Friedman, "Dancing Alone," *New York Times,* May 13, 2004, 25.

35. Thomas L. Friedman, "Too Much Pork and Too Little Sugar," *New York Times,* August 5, 2005.

36. Petrobras production figures. Available: http://www2.petrobras.com.br/portal/frame_ri.asp?pagina=/ri/ing/index.asp&lang=en&area=ri. Brazil's ethanol production in 2005 was about 5 billion gallons.

37. Friedman, *The World Is Flat*, 502.

38. EIA data. Available: http://www.eia.doe.gov/cneaf/electricity/epa/epat6p3.html.

CHAPTER 18

1. Arabian Petrochemical is part of Saudi Basic Industries Corporation.

2. His full name: Saud bin Abdullah bin Thenayan al-Saud. In addition to being the chairman of SABIC, he is also the chairman of the Royal Commission for Jubail and Yanbu and the chairman of the Power and Utility Company for Jubail and Yanbu.

3. http://money.cnn.com/magazines/fortune/global500/2006/snapshots/4091.html.

4. http://www.whitehouse.gov/stateoftheunion/2006.

5. Rachel Layne and Sean Cronin, "Saudi Basic to Buy GE Plastics Unit for $11.6 Billion," Bloomberg, May 21, 2007. Available: http://www.bloomberg.com/apps/news?pid=20601103&sid=a66TqOQoyF14&refer=us.

6. http://en.wikipedia.org/wiki/Kingdom_Centre.

7. OPEC. Available: http://www.opec.org/home/PowerPoint/Downstream%20Constraints/OPECDownstreamexpplans.htm.

CHAPTER 19

1. Yadullah Ijtehadi, "Lack of Office Space Sees Dubai Shoot Up Rental List," *Emirates Today*, July 3, 2006.

2. EIA data. Available: http://www.eia.doe.gov/emeu/cabs/UAE/Oil.html.

3. Organization of Arab Petroleum Exporting Countries, op. cit., 66.

4. Ibid., 75.

5. Part of this growth is due to Dubai's abuse of immigrant labor. In November 2006, Human Rights Watch slammed the UAE for its abuses of construction workers. The abuses allegedly include unpaid or extremely low wages, as well as the withholding of employees' passports and hazardous working conditions. For more, see Human Rights Watch. Available: http://hrw.org/reports/2006/uae1106/uae1106web.pdf.

6. Sonya Crawford, "Halliburton Moves Its Headquarters Abroad," *ABC News*, March 11, 2007. Available: http://abcnews.go.com/WNT/Business/story?id=2942429&page=1.

7. Glenn Thrush, "Hillary Blasts Halliburton's Move to Dubai," *Newsday*, March 12, 2007.

8. Halliburton, "Halliburton Opens Corporate Headquarters in the United Arab Emirates," March 11, 2007. Available: http://www.halliburton.com/default/main/halliburton/eng/news/source_files/news.jsp?newsurl=/default/main/halliburton/eng/news/source_files/press_release/2007/corpnws_031107.html.

9. Michelle Tsai, "Halliburton Says Salaam," *Slate,* March 12, 2006. Available: http://www.slate.com/id/2161652.

CHAPTER 20

1. EIA data show that in 2004, Iran was producing about 3 Tcf of gas per year. Available: http://www.eia.doe.gov/emeu/cabs/Iran/Full.html.

2. BP, "Statistical Review of World Energy 2006," 22.

3. Andreas R. Milhailescu, "Iran Plans to Cut Gas Imports, Subsidies," *Washington Times,* July 3, 2006. Available: http://www.washingtontimes.com.

4. Reuters, "Iran Signs $16 Billion Prelim LNG Deal with Malaysia," January 7, 2007.

5. BP, "Statistical Review of World Energy," June 2005, 20.

6. Author interview with Tripathi, Johannesburg, September 2005.

7. Reuters, "Shell Sees Dilemma over Iran Investment," February 1, 2007.

8. Upstreamonline.com, "Iran 'Strikes $16bn Pars Deal,'" December 20, 2006. Available: http://www.upstreamonline.com/live/middle_east/article125140.ece.

9. *Energy Tribune,* "Sinopec's Huge Iran Deal," January 2007, 28.

10. M. K. Bhadrakumar, "India Finds a $40bn Friend in Iran," *Asia Times,* January 11, 2005. Available: http://www.atimes.com/atimes/South_Asia/GA11Df07.html.

11. Dow Jones Newswires, "Petrobras to Invest $470M in Iran's Caspian Sea," March 6, 2007. Available: http://www.rigzone.com/news/article.asp?a_id =42182.

12. Weekly Compilation of Presidential Documents, "The President's News Conference with President Luiz Inacio Lula da Silva of Brazil at Camp David, Maryland," April 9, 2007.

13. Spiegel Online, "Washington Points Finger at Austrian Energy Firm," April 24, 2007. Available: http://www.spiegel.de/international/business/0,1518,479050,00 .html.

14. Bloomberg, "$3b Investment to Iranian Gas Field from Turkey," September 6, 2007. Available: http://www.turkishweekly.net/news.php?id=48218.

15. EIA data. Available: http://www.eia.doe.gov/emeu/cabs/Iran/Oil.html.

16. Lisa Myers, "Halliburton Operates in Iran despite Sanctions," *NBC News,* March 7, 2005. Available: http://www.msnbc.msn.com/id/7119752. See also Jefferson Morley, "Halliburton Doing Business with the 'Axis of Evil,'" *Washington Post,* February 3, 2005. Available: http://www.washingtonpost.com/wp-dyn/articles/A58298-2005Feb2.html.

17. Melissa Norcross, e-mail statement, from Halliburton media office, February 7, 2007.

18. Halliburton, "Halliburton Completes Work in Iran," April 9, 2007. Available: http://www.halliburton.com/default/main/halliburton/eng/news/source _files/news.jsp?newsurl=/default/main/halliburton/eng/news/source_files/press_release/2007/corpnws_040907.html.

CHAPTER 21

1. During that same period, nuclear power output jumped by 30 percent and electricity generation from all renewable sources increased by 39 percent. Calculations based on EIA data from *Annual Energy Outlook 2006,* Table 5. Excel data available: http://www.eia.doe.gov/oiaf/archive/aeo06/excel/figure5_data.xls.

2. Government Accountability Office, "Gasoline Markets: Special Gasoline Blends Reduce Emissions and Improve Air Quality, but Complicate Supply and Contribute to Higher Prices," June 2005, GAO-05-421, ii.

3. EIA data. Available: http://www.eia.doe.gov/emeu/steo/pub/fsheets/real _prices.html. For data in spreadsheet form: http://www.eia.doe.gov/emeu/steo/pub/fsheets/real_prices.xls.

4. EIA data for the third week of August 2007. Available: http://tonto.eia.doe.gov/dnav/pet/hist/mg_rco_usw.htm.

5. Arthur L. Smith and Aliza Fan, "Oil Is a Good Transportation Value," John S. Herold Inc., May 25, 2007, 7.

6. Bureau of Transportation Statistics. Available: http://www.bts.gov/publications/national_transportation_statistics/html/table_03_14.html.

7. Ibid.

8. Robert Bryce, "Fueling Our Pain," *Salon,* October 11, 2005. Available: http://www.robertbryce.com/101105salon5diesel.htm.

9. Government Accountability Office, op. cit., summary page.

10. Danny Hakim, "A Fuel Saving Proposal from Your Automaker: Tax the Gas," *New York Times,* April 18, 2004. Available: http://www.nytimes.com/2004/04/18/business/yourmoney/18fuel.html?ex=1397620800&en=0b2190e25b4e2ad3&ei=5007&partner=USERLAND.

11. John W. Schoen, "What Does Gasoline Cost in Other Countries?" MSNBC.com. Undated. Available: http://www.msnbc.msn.com/id/12452503.

12. International Energy Agency, *Key World Energy Statistics 2006,* 42, 43.

13. Nick Timiraos, "The Gasoline Tax: Should It Rise?" *Wall Street Journal,* August 18–19, 2007, A4.

14. EIA data. Available: http://tonto.eia.doe.gov/dnav/pet/hist/mg_rt_usA.htm.

15. EIA data. Available: http://tonto.eia.doe.gov/dnav/pet/hist/c100000001A.htm.

16. EIA data. Available: http://tonto.eia.doe.gov/dnav/pet/hist/mg_tt_usW.htm. EIA sales data: http://tonto.eia.doe.gov/dnav/pet/hist/c100000001M.htm.

17. New York Times/CBS News Poll, "Views on the Environment," April 27, 2007. Available: http://www.nytimes.com/imagepages/2007/04/27/us/27pollgraphic .ready.html.

18. William Nordhaus, "The Challenge of Global Warming: Economic Models and Environmental Policy," 27. Available: http://nordhaus.econ.yale.edu/dice _mss_072407_all.pdf. For cost of carbon tax on gasoline, see Daniel Sperling, "A New Carbon Standard," Los Angeles Times, June 21, 2007. Available: http:// www.latimes.com/news/opinion/la-oe-sperling21jun21,0,6045562.story?coll =la-opinion-center.

19. Nordhaus, op. cit., 27.

20. Ibid., 30.

21. Ibid., 27.

22. International Campaign to Ban Landmines. Available: http://www.icbl.org/ treaty/snp.

23. Fatih Birol, "Energy Economics: A Place for Energy Poverty in the Agenda?" Energy Journal, Vol. 28, No. 3, 2007, 3, 4.

24. John Mueller, "A False Sense of Insecurity?" Regulation, Fall 2004, 42–46. Available: http://www.cato.org/pubs/regulation/regv27n3/v27n3-5.pdf.

25. U.S. State Department data. Available: http://www.state.gov/r/pa/prs/ps/ 2006/65422.htm.

26. U.S. State Department data. Available: http://www.state.gov/s/ct/rls/ crt/2006/82739.htm.

27. Those deaths occurred in Afghanistan, Israel, Pakistan, and Thailand.

28. National Weather Service, "Weather Fatalities," undated. Available: http:// www.nws.noaa.gov/om/hazstats.shtml.

29. National Weather Service, "2006 Lightning Fatalities," undated. Available: http://www.nws.noaa.gov/om/hazstats/light06.pdf.

30. Robert Bryce, "Press 0 for Arabic," American Conservative, July 2, 2007. Available: http://www.robertbryce.com/070207TAC-Arabic.htm.

31. Quadrennial Defense Review Report, 2006, 21–22. Available: www .defenselink.mil/qdr.

32. The Central Asia Institute needs money. To donate, go to www.ikat.org, or www.threecupsoftea.com.

33. Greg Mortenson and David Oliver Relin, Three Cups of Tea, 310.

34. National Petroleum Council, "Facing the Hard Truths about Energy," executive summary, July 18, 2007, 2. Available: http://downloads.ConnectLive.com/ events/npc071807/pdf-downloads/Facing_Hard_Truths-Executive_Summary .pdf.

35. *Energy Tribune,* "IEA Revises China's 2Q Oil Demand," April 2007, 26. France has about 112 gigawatts of electric generation capacity. EIA data. Available: http://www.eia.doe.gov/emeu/cabs/France/Full.html.

36. The actual amount of coal-fired electricity was 88.5 percent. About 10.5 percent of the new capacity came from hydroelectric and the remaining 1 percent came from wind. Data via personal communication with Lee Geng, the Beijing correspondent for *Energy Tribune,* April 7, 2007.

37. Reuters, "China faces tight summer power supplies," April 26, 2007.

38. EIA data. Available: http://www.eia.doe.gov/neic/quickfacts/quickelectric .html. In mid-2007, China had about 1.3 billion people. The U.S. had about 300 million.

39. Reuters, "China to Become Top CO_2 Emitter in 2007 or '08—IEA," April 19, 2007. Available: http://www.planetark.com/dailynewsstory.cfm/newsid/ 41461/story.htm.

40. Lee Geng, "Coal Still King," *Energy Tribune,* July 2006, 15.

41. Keith Bradsher and David Barboza, "Clouds from Chinese Coal Cast a Long Shadow," *New York Times,* June 11, 2006.

42. Chris Buckley, "China Says Exports Fuel Greenhouse Gas Emissions," Reuters, June 22, 2007.

43. Some estimates put the added electricity capacity at some 760 gigawatts by 2030. See "Remarks by USAID India Mission Director George Deikun, U.S.-India Energy Efficiency Technology Cooperation Conference, Hotel Maurya Sheraton, New Delhi, May 2–3, 2006." Available: http://www.usaid.gov/in/newsroom/ speeches/may02_6_2.htm.

44. Vinod K. Sharma, "The 'Dirty' Saviour," *Business Standard,* July 7, 2007. Available: http://www.business-standard.com/lifeleisure/storypage.php?leftnm =5&subLeft=2&chklogin=N&autono=290295&tab=r.

45. In 2004, Mexico had installed a generating capacity of about 49.6 GW. EIA data. Available: http://www.eia.doe.gov/emeu/cabs/Mexico/Full.html.

46. Data from India's Ministry of Coal and India's Planning Commission.

47. David Adam, "Flights Reach Record Levels despite Warnings over Climate Change," *The Guardian,* May 9, 2007. Available: http://environment.guardian .co.uk/travel/story/0,,2075294,00.html.

48. Kyle Wingfield, "Europe's Carbon Con Job," *Wall Street Journal,* August 21, 2007, A14.

49. Reports from the European Environment Agency in early 2007 showed that only five of the members of the EU-15 were meeting their carbon reduction targets. For more on this topic, see BBC.com, "In Graphics: The EU and Emissions," January 10, 2007. Available: http://news.bbc.co.uk/2/hi/europe/6244465 .stm.

50. Ibid.

51. Kiichiro Ogawa, "Sustainable Future In Japan—Energy Technology Aspects," *Energy Tribune,* February 2007, 27.

52. *The Economist,* "Powering Up," September 16, 2006, 89.

53. Janet Pelley, "Solar Cells That Harness Infrared Light," *Environmental Science and Technology Online,* March 2, 2005. Available: http://pubs.acs.org/subscribe/journals/esthag-w/2005/mar/tech/jp_solarcells.html.

54. Spectrolab, "Boeing Spectrolab Terrestrial Solar Cell Surpasses 40 Percent Efficiency," December 6, 2006. Available: http://www.spectrolab.com/com/news/news-detail.asp?id=172.

55. Andrew C. Revkin and Matthew L. Wald, "Solar Power Wins Enthusiasts but Not Money," *New York Times,* July 16, 2007.

56. For more on this topic, see the NRC's Web page. Available: http://www.nrc.gov/reading-rm/doc-collections/fact-sheets/new-nuc-plant-des-bg.html.

57. Robert Bryce, "Q & A with Patrick Moore," *Energy Tribune,* June 2006, 9.

58. Richard Simon, "Nuclear Power Enters Global Warming Debate," *Los Angeles Times,* April 9, 2007.

59. Benjamin Grove, "Utah Senator: Yucca 'Does Not Make Sense,'" *Las Vegas Sun,* September 21, 2005. Available: http://www.lasvegassun.com/dossier/nuke.

60. Harry Reid, "Yucca Mountain." Available: http://reid.senate.gov/issues/yucca.cfm.

61. Robert L. Bradley, Jr., and Richard W. Fulmer, *Energy,* 27.

62. Michael Briggs, "Widescale Biodiesel Production from Algae," UNH Biodiesel Group, August 2004. Available: http://www.unh.edu/p2/biodiesel/article_alge.html.

63. Mark Clayton, "Algae—Like a Breath Mint for Smokestacks," *USA Today,* January 10, 2006. Available: http://www.usatoday.com/tech/science/2006-01-10-algae-powerplants_x.htm.

64. Mark Shaffer, "Algae Could Be Fuel of the Future," *Arizona Republic,* October 14, 2006. Available: http://www.azcentral.com/arizonarepublic/business/articles/1014biz-algae1014.html.

65. Paul Israel, *Edison,* 410–421.

66. Credit to Vaclav Smil for first suggesting this idea.

67. http://www.charleslindbergh.com/plane/orteig.asp.

68. Leonard David, "SpaceShipOne Wins $10 Million Ansari X Prize in Historic 2nd Trip to Space," Space.com, October 4, 2004. Available: http://www.space.com/missionlaunches/xprize2_success_041004.html.

69. John Tierney, "A Cool $25 Million for a Climate Backup Plan," *New York Times,* February 13, 2007. Available: http://www.nytimes.com/2007/02/13/science/earth/13tier.html?ex=1187928000&en=3be723dcbb74269c&ei=5070.

70. U.S. Department of Interior, "ANWR Oil Reserves Greater than Any State," March 12, 2003. Available: http://www.doi.gov/news/030312.htm.

71. ANWR.org, "Making the Case for ANWR Development," undated. Available: http://www.anwr.org/case.htm.

72. In 2005, the U.S. used 22 Tcf. Source: EIA, *Annual Energy Outlook 2007*, op. cit., 7.

73. Charli Coon, "ANWR: Drilling for Answers," Heritage Foundation, March 14, 2002. Available: http://www.heritage.org/Press/Commentary/ed031402c.cfm.

74. Alaska Coalition. Available: http://www.alaskacoalition.org/Public_lands.htm.

75. Michael Grunwald and Eric Pianin, "Deals to Block Drilling in Everglades, Gulf," *Washington Post*, May 30, 2002, A1. Available: http://www.washingtonpost.com/ac2/wp-dyn/A30379-2002May29?language=printer.

76. In 2005, Florida consumed about 778 billion cubic feet of gas. EIA data. Available: http://tonto.eia.doe.gov/dnav/ng/ng_cons_sum_dcu_SFL_a.htm.

77. *Energy Tribune*, "The Offshore Drilling Mess," May 2006, 6.

78. Handbook of Texas Online. Available: http://www.tsha.utexas.edu/handbook/online/articles/TT/mgt2.html.

79. *Energy Tribune*, "The Offshore Drilling Mess," op. cit.

80. Ibid.

81. David Wood, Saeid Mokhatab, and Michael J. Economides, "Global LNG Trade on the Verge of Huge Expansion," *Energy Tribune*, May 2007, 23.

82. BP, "Statistical Review of World Energy," 2007. Available: www.bp.com.

83. Yoichi Kaya and Keiichi Yokobori, *Environment, Energy, and Economy*, Chapter 13. Available: http://www.unu.edu/unupress/unupbooks/uu17ee/uu17ee0h.htm.

84. Jesse H. Ausubel, "Renewable and Nuclear Heresies," *International Journal Nuclear Governance Economy and Ecology*, Vol. 1, No. 2, 2007, 230. Available: http://phe.rockefeller.edu/docs/HeresiesFinal.pdf.

85. Robert E. Gillon, "Half a Million Gas Wells in the U.S. and Most Are In Decline," John S. Herold Inc., May 11, 2007, 1.

CHAPTER 22

1. Jerry Taylor and Peter Van Doren, "OPEC Is Not in the House," *National Review Online*, May 24, 2007. Available: http://article.nationalreview.com/?q=Yjg2MTA4N2FhYzU2YTIxMDVhMWE0ODZiMzg3OTE3YTk=.

2. According to BP's 2007 "Statistical Review of World Energy," total global energy consumption in 2006 was 10.88 billion tons of oil equivalent, or about 79.7 billion barrels of oil equivalent. A price of $60 per barrel of oil equivalent yields an approximate figure of about $4.8 trillion.

3. Andrew J. Bacevich, "The Islamic Way of War," *American Conservative*, September 11, 2006. Available: http://www.amconmag.com/2006/2006_09_11/cover.html.

4. Thaddeus Herrick, "The Property Report: U.S. Infrastructure Found to be in Disrepair—Higher Taxes Are Forecast to Meet Investment Need, Reconsidering Cities," *Wall Street Journal*, May 9, 2007.

5. U.S. Treasury data. Available: http://www.treasurydirect.gov/NP/BPDLogin ?application=np.

6. U.S. Treasury data. Available: http://www.treasurydirect.gov/govt/reports/ ir/ir_expense.htm.

7. U.S. Office of Management and Budget, Historical Tables, Fiscal Year 2006, Table 4.1, 76. Available: http://www.whitehouse.gov/omb/budget/fy2006/pdf/ hist.pdf.

8. The U.S. operates a secret military base in Ar'ar, Saudi Arabia, about 250 miles southwest of Baghdad.

EPILOGUE TO THE PAPERBACK EDITION

1. Available: http://www.mahalo.com/Paris_hilton_revenge_video.

2. Corinne Alexander and Chris Hurt, "Biofuels and Their Impact on Food Prices," *BioEnergy*, ID–346-W. Available: http://www.ces.purdue.edu/extmedia/ ID/ID–346-W.pdf.

3. Thomas Elam, Coalition for Balanced Food and Fuel Policy, "Biofuel Support Policy Costs to the U.S. Economy," March 24, 2008, 25. Available: http://www.balancedfoodandfuel.org/ht/a/GetDocumentAction/i/10560.

4. Donald Mitchell, World Bank, "A Note on Rising Food Prices," April 8, 2008, 1. Available: http://image.guardian.co.uk/sys-files/Environment/documents/ 2008/07/10/Biofuels.PDF.

5. National Public Radio, "World Bank Chief: Biofuels Boosting Food Prices," April 11, 2008. Available: http://www.npr.org/templates/story/story/php?storyId =89545855.

6. For more, see: http://www.ifpri.org/about/about_menu.asp.

7. Mark W. Rosegrant, "Biofuels and Grain Prices: Impacts and Policy Responses," International Food Policy Research Institute, May 7, 2008. Available: http://www.ifpri.org/pubs/testimony/rosegrant20080507.asp. Rosegrant's full quote: "The increased biofuel demand during the period, compared with previous historical rates of growth, is estimated to have accounted for 30 percent of the increase in weighted average grain prices. Unsurprisingly, the biggest impact was on maize prices, for which increased biofuel demand is estimated to account for 39 percent of the increase in real prices. Increased biofuel demand is estimated to account for 21 percent of the increase in rice prices and 22 percent of the rise in wheat prices."

8. Ibid.

9. Ibid.

10. U.S. Department of Agriculture, "U.S.D.A. Officials Briefing with Reporters on the Case for Food and Fuel," May 19, 2008. Available: http://www.usda.gov/wps/portal/!ut/p/_s.7_0_A/7_0_1OB?contentidonly=true&contentid=2008/05/0130.xml.

11. U.S. Department of Agriculture, Economic Research Service, "Food Security Assessment, 2007," July 2008, 1. Available: http:// www.ers.usda.gov/ Publications/GFA19/GFA19.pdf.

12. U.S. Department of Agriculture, "Global Agricultural Supply and Demand: Factors Contributing to the Recent Increase in Food Commodity Prices," revised July 2008, 6, 14. Available: http://www.ers.usda.gov/Publications/WRS0801/WRS0801.pdf.

13. U.S. Department of Agriculture, *Amber Waves*, September 2007, 39. Available: http://www.ers.usda.gov/AmberWaves/September07/PDF/AW_September07.pdf.

14. U.S. Department of Agriculture, "Long-Term Projections," February 2007, 31. Available: http://www.ers.usda.gov/publications/oce071/oce20071c.pdf.

15. UN Food and Agriculture Organization, "Biofuels: Prospects, Risks and Opportunities," October 2008, 8. Available: http://www.fao.org/docrep/011/i0100e/i0100e00.htm.

16. Quote is available: http://www.agorafinancial.com/5min/fed-alters-growth-outlook-fomc-minutes-decoded-possible-global-famine-a-suprime-proof-market-and-more/.

17. Robert Bryce, "The Ethanol Apologists," Counterpunch.org, April 17, 2008. Available: http://www.counterpunch.org/bryce04172008.html.

18. Renewable Fuels Association, "Ethanol Facts: Food vs. Fuel," viewed September 8, 2008. Available: http://www.ethanolrfa.org/resource/veetc/.

19. Art Jahnke, "Will Ethanol Save Us or Sucker Us," *BU Today*, April 2, 2008. Available: http://www.bu.edu/today/2008/04/01/will-ethanol-save-us-or-sucker-us.

20. Robert Zubrin and Gal Luft, "Food vs. Fuel, a Global Myth," *Chicago Tribune*, May 6, 2008. Available: http://www.ishs.org/news/?p=132.

21. In August 2008, the USDA estimated that American corn production for the year would total 12.3 billion bushels, of which 4.1 billion bushels would be used for ethanol production. See Scott Kilman, "Bumper Harvests Not Enough to Ease Food Costs," *Wall Street Journal*, August 13, 2008, A3. Available: http://online.wsj.com/article/SB121854537937633263.html?mod=hps_us_whats_news. In 2000, corn ethanol producers used about 571 million bushels. For data see the Earth Policy Institute, which reports that in 2000, U.S. corn ethanol consumed 16 million tons of corn. With 35.7 bushels per ton, that equals 571.2 million bushels. Earth Policy data available: http://www.earth-policy.org/Updates/2007/Update63_data2.htm#fig5.

22. U.S. Grains Council data. Available: http://www.grains.org/page.ww ?section=Barley%2C+Corn+%26+Sorghum&name=Corn.

23. Available: http://www.quotationspage.com/quote/282.html.

24. Available: http://www.eia.doe.gov/kids/energyfacts/sources/non-renewable/ refinery.html.

25. Available: http://www.quoteoil.com/oil-barrel.html.

26. Energy Information Administration data. Diesel data available: http:// tonto.eia.doe.gov/dnav/pet/hist/mdiupus2a.htm. Gasoline data available: http:// tonto.eia.doe.gov/dnav/pet/hist/mgfupus2a.htm.

27. Energy Information Administration, *Annual Energy Outlook 2008 With Projections to 2030*, June 2008, Figure 89. Available: http://www.eia.doe.gov/ oiaf/aeo/pdf/0383(2008).pdf.

28. International Energy Agency, "Medium Term Oil Market Report," July 1, 2008, 20.

29. Ibid., 21.

30. Renewable Fuels Association data for July 2008 show that production was 614,000 barrels per day. Available: http://www.ethanolrfa.org/industry/statistics/.

31. Energy Information Administration data from July 2000 and July 2008. Available: http://tonto.eia.doe.gov/dnav/pet/hist/mttupus2m.htm.

32. Energy Information Administration data. Available: http://tonto.eia.doe .gov/dnav/pet/hist/mcrfpus2m.htm.

33. Energy Information Administration data. Available: http://tonto.eia.doe .gov/dnav/pet/hist/mttimus2m.htm.

34. Energy Information Administration data. Available: http://tonto.eia.doe .gov/energy_in_brief/natural_gas_production.cfm.

35. Energy Information Administration data: http://www.eia.doe.gov/pub/ oil_gas/natural_gas/data_publications/advanced_summary/current/adsum.pdf.

36. Sapphire Energy, "Sapphire Energy builds investment syndicate to fund commercialization of Green Crude Production," September 17, 2008. Available: http://www.sapphireenergy.com/press_release/4.

37. *New York Times*, "UK Sets Challenge for Algae Biofuel by 2020," October 23, 2008. Available: http://www.nytimes.com/external/gigaom/2008/10/23/ 23gigaom-uk-sets-challenge-for-algae-biofuel-by-2020-13288.html.

BIBLIOGRAPHY

Bacevich, Andrew J. *The New American Militarism: How Americans Are Seduced by War.* Oxford: Oxford University Press, 2005.

Baker, James A., III, with Thomas M. DeFrank. *Politics of Diplomacy: Revolution, War and Peace, 1989–1992.* New York: G. P. Putnam & Sons, 1995.

Bradley, Robert L., Jr., and Richard W. Fulmer. *Energy: The Master Resource.* Dubuque: Kendall/Hunt Publishing Company, 2004.

Bronson, Rachel. *Thicker than Oil: America's Uneasy Partnership with Saudi Arabia.* New York: Oxford University Press, 2006.

Bryce, Robert. *Pipe Dreams: Greed, Ego, and the Death of Enron.* New York: Public Affairs, 2002.

———. *Cronies: Oil, the Bushes, and the Rise of Texas, America's Superstate.* New York: PublicAffairs, 2004.

Deffeyes, Kenneth S. *Hubbert's Peak: The Impending World Oil Shortage.* Princeton, N.J.: Princeton University Press, 2001.

Economides, Michael J., and Ronald Oligney. *The Color of Oil: The History, the Money and the Politics of the World's Biggest Business.* Katy, TX: Round Oak Publishing, 2000.

Eland, Ivan. *The Empire Has No Clothes: U.S. Foreign Policy Exposed.* Oakland, Calif.: Independent Institute, 2004.

Fischer, Louis. *Oil Imperialism: The International Struggle for Petroleum.* London: George Allen & Unwin Ltd., 1926.

Friedman, Thomas L. *The World Is Flat: A Brief History of the Twenty-First Century* (Release 2.0). New York: Farrar, Straus & Giroux, 2006.

Goralski, Robert, and Russell W. Freeburg. *Oil and War: How the Deadly Struggle for Fuel in WW II Meant Victory or Defeat.* New York: William Morrow & Company, 1987.

Gordon, Michael R., and Bernard F. Trainor. *Cobra II: The Inside Story of the Invasion and Occupation of Iraq.* New York: Pantheon Books, 2006.

Hadar, Leon. *Sandstorm: Policy Failure in the Middle East.* Houndsmills, United Kingdom: Palgrave Macmillan, 2005.

Heinberg, Richard. *The Party's Over: Oil, War and the Fate of Industrial Societies.* Gabriola Island, British Colombia: New Society Publishers, 2003.

Huber, Peter W., and Mark P. Mills. *The Bottomless Well: The Twilight of Fuel, the Virtue of Waste, and Why We Will Never Run Out of Energy.* New York: Basic Books, 2005.

Israel, Paul. *Edison: A Life of Invention.* New York: John Wiley & Sons, 1998.

Kaya, Yoichi, and Keiichi Yokobori. *Environment, Energy, and Economy: Strategies for Sustainability.* Tokyo: United Nations University Press, 1997.

Kinzer, Stephen, *All the Shah's Men: An American Coup and the Roots of Middle East Terror.* Hoboken, N.J.: John Wiley & Sons, 2003.

Klare, Michael. *Blood and Oil: The Dangers and Consequences of America's Growing Dependency on Imported Petroleum.* New York: Metropolitan Books, 2004.

Kunstler, James Howard. *The Long Emergency: Surviving the Converging Catastrophes of the Twenty-First Century.* New York: Atlantic Monthly Press, 2005.

Lawrence, Bruce. *Messages to the World: The Statements of Osama bin Laden.* London: Verso, 2005.

Lieber, James B. *Rats in the Grain: The Dirty Tricks and Trials of Archer Daniels Midland, the Supermarket to the World.* New York: Four Walls Eight Windows, 2000.

Lovins, Amory B., E. Kyle Datta, Odd-Even Bustnes, Jonathan G. Koomey, and Nathan J. Glasgow. *Winning the Oil Endgame: Innovation for Profits, Jobs, and Security.* Snowmass, Colo.: Rocky Mountain Institute, 2004.

Marton, Kati. *A Death in Jerusalem.* New York: Arcade Publishing, 1996.

Mortenson, Greg, and David Oliver Relin. *Three Cups of Tea: One Man's Mission to Fight Terrorism and Build Nations . . . One School at a Time.* New York: Viking, 2006.

Pratt, Joseph A., Tyler Priest, and Christopher A. Castaneda. *Offshore Pioneers: Brown & Root and the History of Offshore Oil and Gas.* Houston, Texas: Gulf Publishing Company, 1997.

Prindle, David E. *Petroleum Politics and the Texas Railroad Commission.* Austin: University of Texas Press, 1981.

Roberts, Paul. *The End of Oil: On the Edge of a Perilous New World.* New York: Houghton Mifflin, 2004.

Sifry, Micah L., and Christopher Cerf. *The Iraq War Reader: History, Documents, Opinions.* New York: Touchstone, 2003.

Simmons, Matthew. *Twilight in the Desert: The Coming Saudi Oil Shock and the World Economy.* Hoboken, N.J.: John Wiley & Sons, 2005.

Smil, Vaclav. *Energy at the Crossroads: Global Perspectives and Uncertainties.* Cambridge: MIT Press, 2003.

Stagliano, Vito A. *A Policy of Discontent: The Making of a National Energy Strategy.* Tulsa, Okla.: PennWell, 2001.

Woodward, Bob. *Plan of Attack.* New York: Simon & Schuster, 2004.

INDEX